AF788317

Natural Gas in the 21st Century

Edited by Badie I. Morsi, Omar M. Basha, Lina Montuori and Manuel Alcázar-Ortega

Published in London, United Kingdom

Natural Gas in the 21st Century
http://dx.doi.org/10.5772/intechopen.1003409
Edited by Badie I. Morsi, Omar M. Basha, Lina Montuori and Manuel Alcázar-Ortega

Contributors
Amir Sepehri, Beatriz Hernandez-Mendez, Calvin Kwesi Gafrey, Christine Chan, Congning Yang, Edith E. Gutierrez-Segura, Edward Agbai, Edward Aigbedion, Ehsan Hakimi, Fengyang Xiong, Hossein Maazallahi, Jia Chen, Juan Carlos Sanchez-Meza, Julianne M. Fernandez, Lili Liu, Lina Montuori, Manuel Alcázar-Ortega, Miguel Balcazar, Mohammad Kamal Ghassem Alaskari, Paitoon Tontiwachwuthikul, Parinaz Soleimani, Rifaat Abdalla, Robert Wilson, Samuel Tetteh-Abaku, Sergio Suarez-Contreras, Shiwei Wu, Shu Jiang, Teerawat Sema, Tianci Li, W. James Hughes, Yifeng Yan, Yiwen (Sophia) Gong, Yunfan Wen

© The Editor(s) and the Author(s) 2025

The rights of the editor(s) and the author(s) have been asserted in accordance with the Copyright, Designs and Patents Act 1988. All rights to the book as a whole are reserved by INTECHOPEN LIMITED. The book as a whole (compilation) cannot be reproduced, distributed or used for commercial or non-commercial purposes without INTECHOPEN LIMITED's written permission. Enquiries concerning the use of the book should be directed to INTECHOPEN LIMITED rights and permissions department (permissions@intechopen.com).

Violations are liable to prosecution under the governing Copyright Law.

(cc) BY

Individual chapters of this publication are distributed under the terms of the Creative Commons Attribution 4.0 License which permits commercial use, distribution and reproduction of the individual chapters, provided the original author(s) and source publication are appropriately acknowledged. If so indicated, certain images may not be included under the Creative Commons license. In such cases users will need to obtain permission from the license holder to reproduce the material. More details and guidelines concerning content reuse and adaptation can be found at http://www.intechopen.com/copyright-policy.html.

Notice

Statements and opinions expressed in the chapters are these of the individual contributors and not necessarily those of the editors or publisher. No responsibility is accepted for the accuracy of information contained in the published chapters. The publisher assumes no responsibility for any damage or injury to persons or property arising out of the use of any materials, instructions, methods or ideas contained in the book.

First published in London, United Kingdom, 2025 by IntechOpen
IntechOpen is the global imprint of INTECHOPEN LIMITED, registered in England and Wales, registration number: 11086078, 167-169 Great Portland Street, London, W1W 5PF, United Kingdom

For EU product safety concerns: IN TECH d.o.o., Prolaz Marije Krucifikse Kozulić 3, 51000 Rijeka, Croatia, info@intechopen.com or visit our website at intechopen.com.

British Library Cataloguing-in-Publication Data
A catalogue record for this book is available from the British Library

Natural Gas in the 21st Century
Edited by Badie I. Morsi, Omar M. Basha, Lina Montuori and Manuel Alcázar-Ortega
p. cm.
Print ISBN 978-0-85466-855-7
Online ISBN 978-0-85466-854-0
eBook (PDF) ISBN 978-0-85466-871-7

If disposing of this product, please recycle the paper responsibly.

We are IntechOpen, the world's leading publisher of Open Access books

Built by scientists, for scientists

7,500+
Open access books available

195,000+
International authors and editors

215M+
Downloads

156
Countries delivered to

Our authors are among the
Top 1%
most cited scientists

12.2%
Contributors from top 500 universities

CLARIVATE ANALYTICS BOOK CITATION INDEX INDEXED

WEB OF SCIENCE™

Selection of our books indexed in the Book Citation Index in Web of Science™ Core Collection (BKCI)

Interested in publishing with us?
Contact book.department@intechopen.com

Numbers displayed above are based on latest data collected.
For more information visit www.intechopen.com

Meet the editors

Professor Morsi is the Director of the Petroleum Engineering Program at the Swanson School of Engineering, University of Pittsburgh, USA. His research focuses on modeling and optimization of industrial processes; design and scaleup of multiphase reactors; Techno-Economic Analysis (TEA) of the CO_2 capture from flue gas, fuel gas, natural gas, and air using chemical or physical solvents; CO_2 sequestration in geological formation; and Enhanced Oil Recovery (EOR) using CO_2. Professor Morsi has authored and co-authored numerous Journal publications (www. pitt. edu/~rapel), and along with his research group, they gave many invited talks and technical presentations at national and international meetings and leading organizations worldwide. He is the Executive Director of the Annual International Pittsburgh Energy and Carbon Management Conference (PECMC) and the Editor-in-Chief of "Fuels."

Dr. Omar M. Basha is the Principal Engineer and Managing Director of ONB Engineering Research and Technical Services, a specialty firm focusing on data-driven technology development and process optimization within the chemical, energy, and manufacturing sectors. His expertise spans process modeling, techno-economic analysis, and multiphase reactor design, focusing on turning innovative concepts into commercially viable solutions. Dr. Basha holds a Ph.D. in Chemical Engineering and is a licensed Professional Engineer (P. E.). His career spans roles in industry, academia and national labs, where he worked on numerous R&D and process development projects centered on clean energy transitions, carbon capture, and sustainable technologies. He has a strong track record of successfully scaling up new technologies from pilot to commercial stages, having led two startups from early development to licensing. He has also published numerous peer-reviewed papers, holds several patents, and regularly consults on complex projects involving hydrocarbon conversion technologies, carbon management, green hydrogen production, and process intensification.

Lina Montuori obtained her Ph.D. in Industrial Engineering and her MBA from the Polytechnic University of Valencia (Spain). Previously, she had received her B. Sc. and M. Sc. cum laude in Industrial Engineering from the University of Naples "Federico II" (Italy) and Executive Master in Digital Marketing and Analytics from the Catholic University of Avila (Spain). She worked at the Power Center for Utility Explorations of the University at Buffalo (USA). She has a strong track record of leading successful projects at top international brands such as Enel, Solaer Group, and Toshiba Mitsubishi-Electric Industrial Corporation, specializing in renewables, smart grids, energy management, and port automation. Currently, she is combining her work with the position of Professor at the Department of Applied Thermodynamics of the Polytechnic University of Valencia. Furthermore, she is a senior researcher at the Energy Engineering Institute (IIE) with multiple relevant publications in high-impact Q1 journals.

Manuel Alcázar-Ortega received his Ph.D. in Industrial Engineering from the Polytechnic University of Valencia (Spain) and a Ph. D. in Electrical Engineering from the University of South Florida (USA). Previously, he had received his B. Sc. and M. Sc. in Industrial Engineering from the University of Castile-La Mancha (Spain) and the Polytechnic University of Valencia, respectively. Currently, he is an Associate Professor and Deputy Director of the Department of Electrical Engineering at the Polytechnic University of Valencia, where he has been linked to the Research University Institute for Energy Engineering from 2004 to the present. There, he carried out research activities through many different international projects in collaboration with network operators, large and medium energy consumers, distributors, energy traders, and regulators in Europe, the United States of America, and other South American countries.

Contents

Preface

Natural gas has become increasingly central to the global energy landscape, significantly influencing both energy markets and environmental sustainability. This edited volume explores natural gas's role in contemporary energy discussions, highlighting challenges, technological innovations, environmental impacts, and future opportunities.

The chapters within this volume address key thematic areas, beginning with the critical issue of methane emissions, exploring current strategies for emission measurement, mitigation techniques, regulatory frameworks, and their broader climate implications. This discussion is complemented by in-depth examinations of unconventional gas resources, including coalbed methane, tight gas, shale gas, gas hydrates, and emerging technologies in hydrogen production.

Further emphasis is placed on sustainable practices within natural gas production, focusing particularly on environmental impacts, strategies for carbon management, and innovative technologies aimed at minimizing ecological footprints. Perspectives on radiological and environmental aspects of shale gas extraction are also critically examined, highlighting the need for rigorous safety standards and environmental baseline studies.

Technological innovation and sustainable development form another critical theme explored within this volume. The authors provide insights into advancements aimed at improving extraction efficiency, reducing environmental harm, and optimizing production processes. Significant attention is given to technologies such as waste heat recovery and carbon capture and strategies to eliminate or significantly reduce environmentally intensive extraction techniques.

The geographical scope of this volume is diverse, with contributors presenting research and insights from various global regions. The evolving role of liquefied petroleum gas (LPG) and renewable BioLPG is discussed, particularly within the contexts of Europe and Africa, examining regulatory developments, technological advancements, market transitions, and their implications for sustainability and clean energy transitions.

Additionally, methodological advancements in petroleum research are explored, highlighting analytical techniques that enhance fuel quality assessment and efficiency in petroleum production processes. The critical role of geographic information systems (GIS) and remote sensing technologies in assessing risks and interdependencies within oil and gas infrastructures is also discussed, underscoring the importance of robust infrastructure management.

We believe this book will serve as a valuable resource for scholars, industry professionals, policymakers, and anyone interested in understanding the current state and future directions of natural gas utilization in the 21st century.

Finally, we sincerely thank all contributing authors and editors who have diligently worked to create this comprehensive volume.

Badie I. Morsi
Department of Chemical and Petroleum Engineering,
University of Pittsburgh,
Pittsburgh, USA

Omar M. Basha
Principal Engineer and Managing Director,
ONB Engineering Research and Technical Services,
Newark, USA

Lina Montuori and Manuel Alcázar-Ortega
Universitat Politècnica de València (UPV),
Valencia, Spain

Section 1

Methane Emissions and Mitigation Pathways

Chapter 1

Methane Emissions from Natural Gas Systems: A Synthesized Review and Insights into Mitigation Gaps and Pathways

Hossein Maazallahi, Jia Chen and Julianne M. Fernandez

Abstract

Methane is a primary component of natural gas systems, which contributes to supplying global energy demand. This resourceful gas poses high global warming potential and ranks second in increasing global mean temperature after carbon dioxide. While the use of natural gas has been noted as a bridge to move toward renewable energies, exceeding methane emissions from the natural gas supply chain can offset its climate benefits. Measurement-based campaigns have shown that methane emissions from natural gas systems can be mostly larger than what it has been documented, which can be due to the use of generic emission factors, unaccounted emissions and emissions from abandoned wells collectively. Here, we explore into the details of studies, show the effectiveness of measurement-based campaigns in understanding the actual emissions from activities, solutions that can be implemented or further studies in mitigating emissions and challenges that can slow-down or halt the mitigation process.

Keywords: methane emissions, emission mitigation, emission regulations, emission measurement, climate innovations

1. Introduction

1.1 Methane emissions and global warming

Methane has relatively short-term lifetime spanning around 10–12 years [1, 2], which means that considerable emission reduction of this greenhouse gas on a short-term basis has high potential in decreasing its atmospheric abundance, hence slowing down global warming [3]. Methane has a much stronger radiative forcing, with a mass-to-mass ratio indicating it is over 28 times higher in terms of trapping heat on a 100-year time horizon and becomes 84 times stronger effect on a 20-year time horizon [4–6]. This means that adding 1 kg of methane into atmosphere has a similar warming effect similar to adding more than 84 kg of carbon dioxide. On the other hand, reduction or removal of 1 kg of methane from atmosphere is equivalent of reducing warming effect of more than 84 kg of carbon dioxide during a 20-year time horizon.

Methane is considered the second most potent greenhouse gas after carbon dioxide, but its emission reduction has been seen as an opportunity to open a time window implementing actions to achieve global climate goals. Methane is spewed into atmosphere from several natural and anthropogenic sources. Emission reduction of methane from fossil fuel activities is considered a feasible solution in terms of slowing down global warming on a short-term time horizon. This is due to availability of technologies required to reduce or eliminate emissions at release points.

Within the energy sector there is room for improvement to aline toward greener option. This means that the carbon footprint of the sector should be considerably lower than today's level to meet the goals of the Paris Climate Agreement [7]. In **Figure 1**, it can be observed that the global mean temperature has increased relative to 1940 when global temperature records began. It is widely agreed that anthropogenic greenhouse gas emissions are the primary cause of global warming, and the abundance of atmospheric greenhouse gases can be further increased with extra natural GHG emissions, that is, positive climate feedback.

1.2 Methane emission regulations

Methane regulations are of necessity to move toward emission reduction. One of the concerns is that policies may oversee a large number of emission sources [8] or target the wrong locations. In mid-November 2023, the EU Commission announced the release of the first of its kind regulation dedicated to methane emissions from the energy sector [9], which follows the EU methane strategy [10]. The EU methane regulation took place on the 4th of August 2024 [11]. The EU methane regulation has great importance as the EU is the largest importer of fossil fuels worldwide. According to statistical data in 2022, the EU imports 70%, 97%

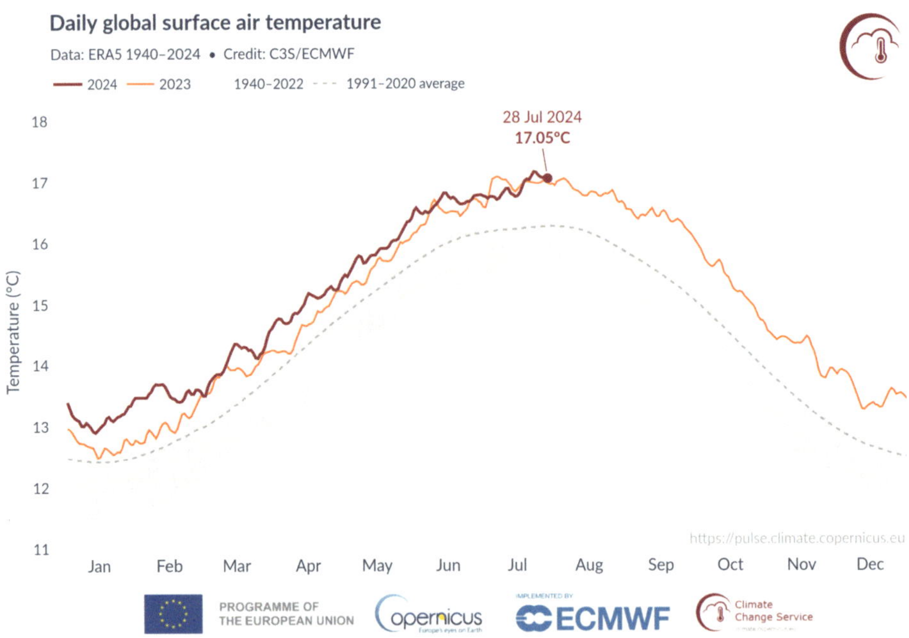

Figure 1.
Global average temperature shows clear increase relative to the previous years since 1940 when the global mean temperature has been measured continuously.

and 90% of its demand for hard coal, oil and natural gas from outside its borders [12], which means that the climate impact of what is used within the European Union happens outside EU. Following the new EU methane regulations, exporting countries of fossil fuel should follow methane regulation to be able to stay in EU market. These regulations are essential in lowering methane emissions which then ask for continuous monitoring through series of measurements for reporting and verification of methane emissions. During the COP26 in 2021, the Global Methane Pledge was announced, which was initiated by the US and European Union and supported by the Global Methane Initiative (GMI), the Global Methane Hub (GMH), the International Energy Agency (IEA), the United Nations Economic Commission for Europe (UNECE) and the World Bank, and 158 countries (until March 2024) have joined the pledge [13].

Stechemesser et al. examined 1500 policies implemented over the past 25 years and identified 63 of the most successful cases [14]. Their study highlights the policies and pathways required to close emission gaps across various economic sectors. Overall, policy implementation should:

i. include a mix of different policies to address emission reduction from multiple perspectives;

ii. differentiate pathways for developing and developed countries, as carbon pricing tends to be more effective in the developed countries, while regulations are more impactful in the developing countries;

iii. demonstrate that successfully implementing these 63 policies could achieve a 26–41% reduction in emissions, contributing significantly to meeting the Paris Agreement targets.

1.3 Methane-related communities, initiatives and work groups

The recent developments in understanding methane emissions from the O&G supply chain and other sources have brought in-depth impacts on real-life conditions. Already, numerous mitigation plans and dialogs worldwide, for example, the recent methane regulations, have been implemented. Several emission mitigations, improvements in technologies, advancement of the market and emergence of science- and tech-based startups have grown. New companies and countries are joining the actions and thinking about how to reduce their carbon footprints.

European Gas Research Group (GERG) is an active R&D group where several methane emission measurements have been tested through coordinated campaigns. From these campaigns, insights have been recommended through the series of measurement-based projects, which have led to enhancement and improvements of measurement techniques, Bottom-Up (BU) and Top-Down (TD) reconciliations at source and site levels [15]. These campaigns were in direct connection with operational industrial activities, and real-world impacts have been obtained to achieve mitigations in real conditions.

The GMH aims to catalyze actions for the agriculture, waste and energy sectors through the global engagement of scientists, industrial experts, policy makers and philanthropists to move toward methane mitigation worldwide [16]. The GMI works on generating energy resources from methane emissions and they provide technical support for these projects [17].

Several standardization groups are active to provide the most recent standards like CEN TC264 WG38 and CEN TC234 WG14. TADI & METEC test centers are the locations where extensive facilities are available to perform methane-emission measurement methods under the controlled operational conditions [18, 19].

International Methane Emissions Observatory (IMEO) is an initiative from UNEP to collect measurement-based ground-based, airborne and satellite data and commission studies and synthesize the data for publications in collaboration with the O&G industry [20]. With regard to the reporting and mitigation purpose of oil, the Oil and Gas Methane Partnership 2.0 (OGMP 2.0) provides a measurement-based framework with mitigation possibilities from which companies can benefit [21]. IMEO initiated the Methane Alert and Response System (MARS) program in which multiple stakeholders contribute to detect and quantify methane emissions from O&G operations. Through the MARS program, governments are informed about large-scale emissions detected through multiple satellites from several organizations or companies. The governments involved in the program are asked to engage, take action and mitigate the detected emissions. With the current consolation of satellite, recent developments and planned launches, this program holds a great position in methane emission mitigation and paves and eases the way for carbon dioxide mitigation.

Methane Mitigation Summits is the largest platform where large events are held to welcome experts from O&G companies to exchange knowledge on how to mitigate and decarbonize O&G industries together with the Industrial Decarbonization network [22]. They have also released a series of podcasts in which different topics have been discussed in detail for the industry and broad audience.

1.4 Top-down/bottom-up methane measurement approaches

There can be two ways to interpret the definition of TD and BU: (I) related to the design of an inventory and (II) related to how measurements are performed.

In the first category, if a technique is applied at all sources within a target area, it is said that the total emissions from the target area are derived from *inventory* BU approach, while if total emissions from that region are derived first, then the total emissions are broken down to the contributors, which can be referred to as *inventory* TD approach.

In the second category, the term *measurement* BU approach refers to a measurement technique in which all emissions from a source are captured and then, they are translated into emission rates. For example, emissions rates derived form an enclosure falls into the *measurement* BU approach. At industry use of generic factors to estimate emission rates, using handheld instruments is also considered a *measurement* BU approach. While the term *measurement* TD approach refers to a measurement in which behaviors of an emission plume in the ambient air is studied through series of measurements and then emissions rates can be derived. For example, mass balance measurements using a plane is considered a *measurement* TD approach. Detection and quantification performance of these techniques can be intercompared [23] as well as for instruments [24]. In a summary, the BU techniques are those that are based on the direct measurements for detection and quantification or only detection of methane emissions and the TD techniques are based on translation of ambient methane abundance concentrations (mixing ratio) into emission rates.

There can be tens, hundreds, or thousands of methane emission points at a site, and all these sources need to be taken into account to obtain total emissions from a target area, which can be accomplished by applying measurement-based techniques.

There are several BU and TD methane emission measurement techniques at source or site levels [25].

The measurement-based TD and BU methods each have specific strengths and weaknesses [25–27].

- TD methods can detect and quantify total emissions from a source relatively faster than BU methods.
- BU approaches can be more sensitive to smaller emitters, such as using bagging or chamber measurements, whereas TD methods generally have higher detection and quantification thresholds.
- TD methods, such as drone-, vehicle-, airborne- and satellite-based techniques, can cover larger areas for identifying sources.
- TD approaches can quantify emissions from large facilities but may not necessarily identify the root cause of the emissions. In such cases, applying BU approaches or using OGI/QOGI cameras can be beneficial.

The applicability of TD methods may be limited by environmental factors, such as wind speed and direction, whereas BU methods do not face these challenges.

1.5 Methane emissions from natural gas systems

Natural gas as a fossil fuel resource contributes significantly to supplying global energy demand. While traditionally, this source has been considered a clean energy source relative to oil and coal due to lower carbon dioxide emissions to produce a unit of energy, intentional and unintentional methane emissions from natural gas systems put this under question. Natural gas remains a significant source of energy for the EU [28]. In 1980, the US Department of Energy was investigating new sources of natural gases from conventional or unconventional sources [29], and exploration of new sources has continued ever since. It is possible that natural gas will be a supporting natural source of energy to supply energy demand due to its low cost [30].

Reducing anthropogenic methane emissions brings financial, environmental and health benefits while it is also important in keeping operation in natural gas systems safe. While minimizing methane loss provides direct financial benefits [31], the other mentioned advantages are sources of indirect financial gains. Emission reductions and decarbonization from the global natural gas industry require visionary and deep efforts to achieve meaningful progress toward emissions reduction.

Methane emissions in natural gas-producing countries showed a high contribution to the total anthropogenic emissions within a country. This presents opportunities that are economically beneficial, in line with regulations within a country or region, environmentally friendly, and can be partially beneficial for supplying demand, especially for countries short of energy. Natural gas systems contribute about 40% to total anthropogenic emissions in the US [32]. This is while in Iran, a country with the second proven natural gas reserve, methane emissions from the energy sector contribute 60% to total anthropogenic emissions [33].

Methane emissions in the natural gas industry are heavy-tailed, which means that there are locations with extremely high levels of methane emissions, which account for the majority of the emissions. This can be observed in all streams of the natural gas industry [30, 34].

1.6 Inventoried methane emissions

Inventories are based on generic emission factors and give an estimate of emissions over a year. However, emission measurements have shown that generally inventories underestimate emissions. This can be due to several reasons including episodic emissions that are most likely not included in the inventory estimates. Estimates of blowouts may not also be included in the inventories due to safety issues. In these events, emission control and repair hold a higher priority than the estimation of emission measurements. The use of generic factors for estimating methane emissions from activities in the natural gas system is another reason. Unreported emissions due to overseeing the emissions in remote areas or due to lower priority relative to safety issues during repair are another reason [35–38].

Methane is the major part of natural gas and the main source of energy within the natural gas system. While methane has large financial and energy-related benefits, intentional or unintentional releases of this gas into the atmosphere from natural gas systems impose threats to achieving climate goals [8]. Among other fossil fuel sources, the use of natural gas to supply demands is considered the cleanest in terms of carbon dioxide emissions to produce a unit of energy relative to dependency on oil or coal, which depends on the anthropogenic emission throughout the whole natural gas system [34, 39]. If methane emissions exceed more than 1–5% of the production across the supply chain, this offsets the climate benefits of the use of natural gas [40]. Normalized methane emissions by production from natural gas systems, that is, the amount of methane emission per unit of production, is very debatable, which can be due to different operational conditions through studies, selection of studied areas, different upscaling methods, and different production rates during studies. The TD studies in the US have shown the normalized methane emissions by production to be greater than 1%, and this normalization can be highly greater for conventional production sites (>10%) relative to unconventional production sites for specific regions, which might be explained by use of newer technologies [41, 42]. This also shows that the normalized methane emissions can be spatially very different. Production normalized can be 6.2–11.7% in the western natural gas field in the US [43]. It was also reported that natural gas production from both conventional and unconventional sources can have a worse environmental impact compared to coal or oil [44]. While natural gas has been considered a bridge to achieve the transition period to move toward a sustainable supply of energy demand [45], the high rate of methane emission from fossil fuels imposes questions about this.

In 1992, the US Environmental Protection Agency (EPA) and Gas Research Institute estimated that 6.04 ± 2.01 Tg was emitted from all natural gas operations in the US, which was equivalent to 1.4 ± 0.5% of total production [46]. In another study, methane emissions from natural gas systems were estimated to be 4.4 Tg $yr.^{-1}$ [47]. Due to inaccurate assumptions or poor-quality data or a combination of both, the replacement of coal and oil with natural gas could not be defended [48]. In 2017, in a synthesis article, methane emissions from extraction to delivery were estimated to be 1.7% (95% CI 1.3% – 2.2%). However, several studies have shown that actual methane emissions from natural gas systems can be significantly different, mostly higher than what is reported in the inventories, and for individual sources can be more uncertain [37].

In 1999, with scaled-up analyses of methane emission measurement from some natural gas production and transmission operations in Russia, it was concluded that the total methane loss from the entire natural gas system in the country was 1.5–1.8%

(4 Tg $yr.^{-1}$), which was lower than other studies [49]. In another study, methane emission loss from natural gas systems in Russia was estimated to be around 25–35 Tg $yr.^{-1}$ [3], which is significantly larger than the other study and can be translated to 9–16% methane loss from the natural gas system in Russia. In 2005, authors of the study from 1999 indicated that methane emission from the largest natural gas transmission pipeline in Russia is 1.4% of transported gas and suggested that leakage reduction of natural gas could be useful for climate change mitigation in the short term compared to coal and oil [50]. Although in Britain it was aimed to keep methane emissions from natural gas systems at 1% supply, a study from 1990 concluded that methane emissions from natural gas systems in the UK were more than 1.9% and most likely between 5.3 and 10.8% [51]. With limited measurements and upscaling to global urban areas, cities can contribute to 7–15% of global anthropogenic methane emissions [52].

As countries are committing to emission reduction from their activities, it is essential to perform accurate measurement techniques to be able to follow the reduction efforts [35], and this falls into the measurement, monitoring, reporting and verification system.

1.7 Methane emission measurements

Methane emission can be measured through a series of detection, quantification and attribution techniques [53], which is going to be expressed in details in Section 2. Depending on meteorological conditions, emission source properties and suitable measurement-specific conditions, it can or cannot be possible to translate detected emissions to source emission strength. This means that to quantify detected methane emissions, certain criteria should be met. Emissions can be quantified using ground-based, aerial, or satellite platforms with *in situ*, open-path or remote sensing techniques. Quantification methods can be based on theoretical equations such as the Gaussian plume model or empirical equations, which are based on several controlled releases. To properly quantify methane emissions, TD and BU methods, explained later in Section 2, are required to be applied to make sure no source is missed within a small or large area [27]. Attribution techniques are those that are used to identify types of emission sources, that is, thermogenic, microbial and pyrogenic. These techniques can be based on the evaluation of concurrent species emissions, isotopic measurements or the use of spatial data to understand an emission point with additional data like wind direction. These techniques can be further used to understand the contribution of different sources in a plume, which refer to the apportionment of source emissions.

All in all, the efforts to frame actual methane emissions from the sources should finally lead to a decrease in its atmospheric abundance. These efforts can be categorized into emission avoidance, destruction, capture, sequestration and utilization of this gas. Emission avoidance and destruction at the release points are the efforts that can be made to directly reduce emissions from activities.

1.8 Methane emissions avoidance and destruction (emission reduction), capture, sequestration and utilization

Rising atmospheric methane abundance imposes strong challenges in achieving the goals of the Paris Agreement [54], and rapid and impactful actions are required to reduce and eliminate the additional atmospheric abundance of this molecule. These efforts can be in reducing emissions from sources through (I) emission avoidance actions like leak repairs, (II) stoppage of flaring or producing products from the flare gas instead of spewing into the atmosphere, (III) destructing the emissions at the release points and (IV) capturing atmospheric methane for requestion and utilization activities. The latter is currently in the research phase.

Methane emissions can be either reduced through avoidance through the repair of emissions points or by destruction if engineering procedures allow. There can be several sources of methane emissions from natural gas systems throughout the supply chain, for example, at wellheads, during transmission lines, at venting stacks, leaks at urban gas networks. The practices, together or individually, can result in methane emission reduction throughout the supply chain.

Methane can be released from natural gas systems upstream, midstream and downstream of the supply chain. Contribution of methane emissions from different parts of these systems depends on a variety of parameters, including infrastructure age, material, maintenance, etc. Although methane emissions from oil systems significantly contribute to total emissions from fossil fuel exploitations, here the focus is on the methane emissions from natural gas systems.

2. Methane emission measurement methods

There is not a single magical measurement solution. There can be a large number of traditional and innovative solutions that can be offered to measure from an emission source [27]. Because currently there are a large number of technologies in instrumentation, different platforms that can be used for mounting the instruments, and a wide range of emission sources exist at a variety of spatial scales, geographic and physical conditions [53]. It is important to know that a single method is not enough and most of the times integration of several methods brings best solution in our understanding of emission sources.

Currently, it is widely accepted that direct methane emission measurements are the best practice to enlighten the amount of methane released into the atmosphere and speed up the mitigation process. It is also widely accepted at the industry, academia and policy levels that to accelerate decarbonization from the energy sector, which includes natural gas operations, and emissions from the bigger emitters should be reduced or eliminated first in the best practice. Hence, detection and quantification of methane emissions from the gas leaks in the natural gas systems should be implemented to properly inventory the leak locations and their emission strengths for mitigation. In the later part of this chapter, mitigations will be discussed where mainly two practices are introduced: (I) emissions avoidance and (II) emissions destruction. Methane capture can be of use later, which is currently in the research phase with challenges to realistically provide large-scale methane capture either at the emitting source or from ambient air.

Selection of methane emission measurement methods, platforms and instruments is highly and mainly dependent on the (I) geospatial location and (II) physical conditions of the emitting target, (III) meteorological conditions and (IV) first-order estimate of the target's emission strength.

i. *Geospatial location:* Some emitting targets are hard to access or receive flight permits for the implementation of ground- or airborne-based techniques. It is possible that data retrieval from satellite may not possible, due to location of sources, for example, offshore, or cloud covers.

ii. *Physical conditions:* Although some methods are applicable to be performed for one source, those are not suitable for another one. For example, one method can be applied for a vent stack, while it is not possible to implement the same method for gas leaks from natural gas distribution networks in urban areas.

iii. *Meteorological conditions:* Some methods are limited to meteorological conditions including wind conditions and wind speed. For example, during unfavorable wind direction, a measurement platform cannot detect any signal from an emitting point. Moreover, low wind or very high wind speeds can be unfavorable for some techniques or result in high emission quantification uncertainties.

iv. *First-order estimate of target's emission strength:* Some techniques are not capable of detecting and/or quantifying certain emitting targets if they are below detection and/or quantifying limits. This also depends on the distance and sometimes on the meteorological conditions. A strong source may not be detectable from a large distance due to large atmospheric dilution between the source and measurement point. A source may be detectable under certain wind conditions but it cannot be quantified if for example wind speed is less than a certain limit (e.g., 2 m s^{-1}) for TD quantifying methods.

2.1 Methane emission measurements at various spatial scales

Measurement techniques can be applied at the source-, site- or regional level. Site-level emissions are defined as total emissions from all of the source-level emitting targets in terms of number of sources and emission strength. Similarly, regional level refers to normally a large region, in scale of km * km, which includes several sites or tens of wellheads. At a very large scale, the regional level can be referred to as emissions at the size of a country.

A natural gas wellhead, a separator, collection points, transmission and distribution lines, compressor stations, and a tank are examples of emitting sources. A petrochemical plant or refinery can be considered a bigger site in which tens, hundreds or thousands of sources can exist. A hydrocarbon basin can be considered a very large-scale emitting target.

2.2 Leak detection and repair (LDAR) program

Historically, natural gas companies are required to monitor gas leaks in urban areas to ensure safe operations. With new regulations, such as the EU methane regulations, these companies who operate in the EU or export natural gas to the EU are required to perform quantification of methane emission within their operation, meet certain criteria, followed by emission reduction over time. For example, the Leak Detection and Repair program [55] is a widely used term among natural gas companies to monitor sites or natural gas distribution networks to detect emission points and repair the loss, which results in lowering emissions over time. The overall emissions from data collected based on the LDAR system can be estimated using generic factors. Although there are several criticisms about the use of the generic factors, which can result in underestimation of overall emissions reported in inventories, relative emission estimates based on these generic factors showed a decrease in emissions after the implementation of the LDAR program [56]. Flame Ionization Detector (FID) or Optical Gas Imaging (OGI) cameras are those commonly used devices in the LDAR program. FIDs, always connected to a probe, can be used directly next to possible emission points in a facility and the operators who hold the FIDs with the probe can walk over buried pipelines. The latter is known as the carpet method. Recently, new technologies have been integrated within the OGI cameras and transformed them into Quantitative Gas Imaging (QOGI) cameras. These cameras can be operated either as handheld or stationary for continuous monitoring. In addition to the LDAR program, if certain criteria are met, the QOGI cameras can also be used for quantification purposes. Not to mention that trained canines were/are used to detect and localize emission points.

The canines are trained to find emission points of odorants added to natural gas, which are also the locations of methane-emitting points. This can be considered as an *indirect methane emission localization practice*. The manufactured instruments are either based on the detection of a whole range of Volatile Organic Compounds (VOCs) or methane-specific. The VOC-specific devices can be used to detect methane emission points knowing that methane is also included in the composition. These are devices to be used for *semi-direct methane emission measurements*. The methane-specific instruments, *in situ* or optical or other technologies, provide *direct methane emission measurements*.

2.3 Measurement instruments and platforms for emission detection and quantification

Traditionally, natural gas leaks in facilities were detected by humans using, for example, handheld FIDs, optical imaging, acoustic leak detection or soap bubble detection. These methods are still in use. However, automatic leak detection operations have been in operation recently, for example, with the use of stationary *in situ* measurements or optical surveys with backend algorithms based on artificial intelligence and machine learning [57, 58]. Automatic gas leak detection can also be based on an automatic drone-based platform [59].

Instruments can be installed onboard vehicles, ships, drones, helicopters, planes, or satellites or used as handheld or on poles, tripods, etc. Basically, with current technologies, depending on (I) the type of the instrument, (II) the way those are designed and used and (III) the type of emission sources, a variety of platforms can be used.

Among the instruments and platforms, stationary instruments (e.g., fixed sensors, ground-based remote sensing, OGIs and QOGIs) are considered as tools for *continuous* monitoring, and relatively speaking, satellites are considered for *semi-continuous* monitoring, daily, weekly or monthly depending on satellites' temporal resolutions. It is worth noting that the OGI and QOGI cameras can be also used as an handheld instrument and are not always stationary. More frequent satellite-based data will be publicly and/or commercially available with current and planned regional and target-based satellites. With current technologies, Landsat 8 [60], Sentinle 2B [61], TROPOMI [62, 63], PRISMA [64], GHGsat constellation satellites [65], MethaneSat [66] and Carbon Mapper [67] are leading satellites in terms of space-borne measurements. These satellites have different revisit times, spatial resolutions, detection and quantification thresholds. However, they can be used solely or together to map and quantify emission points, areas or regions.

Vehicles, drones, helicopters, planes and ships are generally used for detection and quantification purposes. These methods generally have better detection and quantification thresholds, are more precise and require a higher labor force relative to stationary and satellite-based platforms. Not to repeat the detection and quantification threshold is highly dependent on several factors, including distance and strength of emission points, etc.

2.3.1 Vehicle-based measurements

Vehicle-based methods are mostly considered *measurement* TD approaches. Among these techniques, it is possible to mention the tracer method [68, 69], the Gaussian plume dispersion method [70], Other Test Method 33a (OTM-33a) [70, 71], urban leak detection and quantification methods [72, 73]. These methods can be performed in motion, that is, mobile methods, like urban leak detection and quantification method, or stationary like OTM-33a, or performed in either mobile or stationary depending on the situation like the tracer method [74]. Trucks can also be used to measure emissions where larger labs are required to move from one location to another to perform either mobile or stationary measurements. For example, the tracer method, Differential Absorption Lidar (DIAL) method [75], or suction method [76] can be performed using trucks. While the first two truck-based methods are considered a *measurement* TD approach, the latter is a *measurement* BU approach as it is designed to fully capture emissions of gas leaks from a buried pipeline and translate them to emission rates. Some of these methods can also be applied using other platforms, like helicopter-based DIAL measurements [77] or tracer method using drone-based measurements, while the suction method can only be applied using vehicle-based measurements.

2.3.2 Ship-based measurements

Ships are another platform where instruments can be installed and used for offshore measurements. The ships can circle around or zigzag downwind offshore production platforms, which can be used for both detection and quantification purposes using research vessels or cargo ships [78–80]. The use of research vessels has value in designated places that are already available in the vessels, including dry and wet labs but are more expensive to operate compared to cargo ships, which can either be rented for a measurement purpose or can perform measurements and sampling while in normal goods transportation operation. These measurements can be related directly to natural gas operations, or characterizing natural emissions [81, 82], which can possibly have thermogenic sources and increased due to warmer climate, which means positive climate feedback, as with higher atmospheric temperature (**Figure 1**) results in higher water body temperature, and hence increase in seabed methane flux. The ship-based measurement methods are within the *measurement* TD approach. These methods are useful for methane emission characterization from offshore natural gas activities and methane emissions from natural sources.

2.3.3 Airborne measurements

Airborne methane emission measurements include the use of drones [25, 83–86], helicopters [25, 77, 87, 88] and aircraft [36, 89–93] in detecting and quantifying the emissions from offshore and onshore sources. These techniques can be applied to various sources and not only to natural gas systems as the core basics of the techniques are the same and adjustable. Further developments, like the automated pipeline gas leak detection concept, can be an innovative way to save resources [59]. The use of tens or hundreds of drones, swarm drones, in the process of *in situ* or open-path measurements can be another innovative idea to quantify large-sized and/or large-scale sources. The airborne measurements are considered within the *measurement* TD approach. The airborne measurements are useful for the application of emission characterization for both onshore and offshore sources. Drones have less battery span relative to flights of helicopters and aircraft, but drones can fly closer to the sources and be used for smaller sources. The helicopter and airborne platforms have the advantage in carrying heavier and more precise instruments.

2.3.4 Satellite measurements

Regional methane emissions can be studied using satellites with large spatial resolutions like Sentinel-5P/TROPOMI, which can be used for emission measurement of large areas [94, 95] or sources with very large emission rates [96]. Target-focused satellites can be implemented in detecting and quantifying methane emissions from sources in smaller spatial scales within a region onshore and offshore [97–100]. The satellites with large spatial resolutions can be integrated with the target-based satellites to confine emissions sources and their temporal resolutions. The combination of large-scale sensors with target-based satellites can be used to understand methane emissions from a source from the beginning until it is fixed or repaired below the emission detection threshold of satellite sensors [96]. Another advantage of these combinations is the availability of data from several sensors to cover more days when the emission is happening as the satellites may not pass on the same spot every day or the data do not pass quality checks due to, for example, cloud covers.

Currently, satellite measurements from a variety of sensors are used for the individuals and stakeholders who have possibilities in addressing the emission reduction efforts through the MARS program *via* the IMEO data portal [101]. Moreover, for research purposes, commercial satellite imageries can be accessible to researchers through the European Space Agency (ESA) website [102].

Satellite imageries generate large amounts of data continuously, and with the developments of the next generation of satellites, the database becomes extraordinarily large. This calls for the automatic detection of plumes [103], which can be further studied in detail and integrated with ground-based measurements.

2.3.5 Ground-based remote sensing: Total column measurements

Total column measurements are another type of technique that is used for quantifying total emissions from a strong source [104–106] or large cities [107, 108] or basin. Munich Column Carbon Monitoring network (MUCCnet) is an integrated autonomous monitoring network in which total column measurements play a crucial role [109]. The total column measurements can be used for assessing methane and associated emissions like ethane from oil and gas activities [110]. Mobile *in situ* measurements together with total column measurements can be used to quantify, attribute and geo-locate emissions from large festivals in which most of the emissions could be attributed to incomplete combustion and leaks of natural gas within the boundary of the Oktoberfest festival [111, 112]. Ground-based remote sensing can also be used to validate satellite-measured concentration gradients [113] and to better understand the accuracy of emission assessments using satellites.

2.4 Emission attribution and apportionment methods

Naturally, methane molecules are produced in several processes. How a molecule of methane is formed leaves a signature on the molecule. These signatures can be generally classified into three genetic categories: (a) biogenic, (b) thermogenic and (c) pyrogenic. Finding out the origin of methane molecule formation and its process is called methane emission attribution. With the determination of this signature, we can understand how the molecule was generated and, therefore, which sector the emissions belong to. The attribution process can be accomplished mainly with three methods individually or together to gain higher certainty and/or confirmation in the attribution process: (I) measurements of methane isotopes: carbon, $\delta^{13}C$, and/or hydrogen, deuterium δD, (II) determination of presence or absence of certain other molecules in methane plumes and (III) use of geographical information.

2.4.1 Use of isotopes

Wet natural gas, accompanied with higher C_{2+} contents (hydrocarbons with multiple carbon atoms), intends to be depleted in δD ranging from −260 to −150%, while dry natural gas is enriched in deuterium (−180 to −130%.) [114]. While separation of genetic signatures may not be possible in a single graph, $\delta^{13}C\delta D$ graph has shown separation between different samples in several studies [115].

2.4.2 Presence or absence of other gases

Ethane is the second most abundant gas in natural gas, which is normally measured to attribute detected methane plumes to emissions from natural gas systems. While at the production site, the ethane: methane (C_2:C_1) ratio is not very fixed, in natural gas distribution networks in urban areas the C_2:C_1 ratio is usually less than 10%. The natural gas distributors should abide by national standards in which the range of this ratio is determined and varies from country to country.

While natural gas distribution leaks can be detected by simultaneous measurements of methane and ethane, incomplete natural gas combustion can result in the presence of methane and ethane together with carbon dioxide [116], which can be a way to attribute emissions from incomplete combustion sources [117].

2.4.3 Use of geographical information

There are cases where several different sources, either within the same or different genetic categories, can be spatially geolocated.

In the most convenient case, a single source exists, which can be considered the methane emission source for the detected plumes, for example, a single natural gas wellhead in a remote area. In this case, the attribution does not necessarily require measurements of isotopes or possible concurrent gases.

In a more complex case, several sources are present but within the same genetic source, for example, several natural gas wellheads in a gas park. In this case, we know that all emissions are thermogenic but can possibly be from several sources, which require further investigation to properly localize emission source (s) within a cluster.

In the most complicated case, there are several sources from different genetic classes that are spatially clustered in a close spatial proximity. In this case, the attribution of the plume(s) detected in *measurement* TD surveys required the integration of several analyses. For example, if a natural gas wellhead is in close proximity to a biomethane site and an agricultural farmland, plumes detected in *measurement* TD surveys may require analyses of, for example, wind direction, isotopes, presence, or absence of other gases. For example, isotopic analyses from airborne measurements give mean representative values of all emitting sources in a large region, which can be used to find out the contribution of sources to the total emission, that is, emission apportionment [118].

The geographical locations of emitting sources can be based on inventories, field observations or satellite/airborne optical imageries. It is possible that the emissions from production platforms are not included in inventories [79]. The locations derived from satellite or airborne imageries can be either recognized manually through eyeball observations or through deep machine learning [119]. The dataset derived from satellite or airborne imageries through machine learning can be useful for recognizing well pads or facilities that generate a pattern to detect. Moreover, there can be millions of abandoned natural gas and oil wells in a country, which are not pinned or, at the moment, not visible remotely due to natural coverage. These abandoned wells and belonging facilities are hard to find even in on-ground surveys. It is reported that the abandoned wells can be magnetically located using magnetometer of a smartphone and drone [120].

2.5 Measurement uncertainty

Uncertainty associated with measurements holds no less value than measurement itself. Sources of uncertainties are required to be accounted for properly. Quantification methods should be applied during favorable conditions, proper instruments should be used, operators need to pay attention to the details and several measurements can be applied to reduce the overall uncertainty. With current technologies, the uncertainty of emissions quantification is less or the least influenced by the instrument precisions. Controlled situations can be studied to understand the sources of uncertainties and reduce or eliminate them to obtain the best measurement practice in real-world conditions [121], as well as to introduce new methods to quantify uncertainty for very low-emitting sources [122]. Furthermore, the measurement uncertainties can be reduced with the use of machine learning techniques [123]. Controlled release experiments are another way to investigate emission quantification uncertainties [124]. The way, a controlled release experiment is conducted should be adjusted based on the instruments used, platforms instruments are mounted on and sources from which further investigations are meant to be quantified.

2.6 Measurement campaign planning

Methane emissions from natural gas systems can occur onshore and offshore from all streams of the system. This brings a large number of conditions, uncertainty, logistics, safety considerations, instrument availabilities and capability checks, personnel readiness, measurement deployment permissions (if required), etc., to properly implement a successful campaign to obtain high-quality field data. There have been several coordinated large and small campaigns to investigate methane emissions from the supply chain. Weather simulations and historical weather data can be used to find the most suitable campaign planning time openings. For large-scale campaigns, the time for planning needs to be large enough as other scheduling and planning are involved. To fine-tune the measurements in the time window determined, the daily preparation for the campaign can then be determined using weather prediction day(s) close to the execution of measurements. Optimization of measurements can also be dependent on other unpredictable or sudden incidents, such as a pandemic outbreak or instrument failure.

Methane emissions can be episodic [36, 125, 126] within the supply chain and to get a good picture of annual emissions from operations, this needs to be accounted for, that is, addressing the source of over/underestimations during snapshot measurements [126]. These snapshot measurements, which are sampled from emitting locations, are used for extrapolations spatially and temporally. It is not only important to sample properly in the first hand in order to get good representativeness of the population, the samplings are required to be extrapolated with the right methods temporally and spatially [127].

Emissions from sources vary significantly over time and space [26]. This variability poses challenges for accurately assembling emission quantifications in national inventories. Incorporating diverse sources into the inventory process helps account for the temporal variability of emissions, which can further explain discrepancies between measurement-based quantifications using TD and BU estimates [128].

3. Emissions from supply chain

Many studies have been performed around the globe to understand methane emissions from the O&G supply chain worldwide [129]. The more measurements performed, the more they enlighten emissions from sources across the supply chain. Methane emissions can be observed from the production sites, including flaring [130–132], natural gas processing and gathering [43, 133–135], distribution-transmissions [136–138], end users mostly in cities from distribution networks [139], cars, homes, regulators, furnaces, etc. [116, 140–151]. All these sources contribute partially to the total emission and relevant mitigations including repair actions can be done. In another study, it has been shown that the low production sites contribute on average about half (37—75%) of the total O&G methane-related emissions in the US, while these sites contribute 6% to the total O&G production in the country [152]. This shows that these sources can provide great opportunities for mitigation points. Although abandoned O&G wells are not contributing to the supply chain, their emissions are unknown or may not be counted in the inventories, and their emissions are the most uncertain source from the past activities and their numbers are increasing over time [153]. This shows that these sources can further increase methane emissions from past activities, and emission mitigation from the abandoned O&G wells can be a target.

Emissions from the supply chain can be either continuous or episodic. Incidents in which a rupture or a blow-out happens are among the episodic emissions, which can have very large emissions in a short time. On the 26th of September 2022, one of the largest incidents happened offshore for the Nordstream pipeline. This incident was in a short time but released a large amount of methane at the time [154–157]. The cumulative sum of these incidents can account for a large contribution to the total emissions from natural gas activities, and it was estimated that the emissions from the Nordstream pipeline incident were 0.1% of all anthropogenic methane emissions in the year 2022 [158, 159]. Meanwhile, it is possible that these incidents may not be accounted for, as repair of the leaks has higher priorities than quantification. The most recent blowouts, which were studied by several satellite observations, show very large emissions (20–50 t-CH_4 hr.$^{-1}$) over the course of 205 days [96].

4. Methane emission mitigation: Avoidance, destruction and/or removal

Companies are acting actively to find the best solution to offset their carbon footprint, in which practical and operational solutions with considerable resources needed are undeniable. In order to decrease the atmospheric methane abundance and decarbonize industries, several pathways can be taken, which fall into (I) emission reduction at the sources, (II) atmospheric methane capture and removal, (III) atmospheric methane destruction and (IV) replace methane-emitting sources with green solutions like transiting from use of fossil fuel to renewable sources in the energy sector. These efforts fall into two main categories: (I) emissions prevention and (II) reduce or eliminate carbon footprints of activities. There have been several long-term and deep discussions and debates on how to proceed productively without using more resources to save less and avoid greenwashing.

Methane emission reduction at the sources can be done, for example, through the repair of leaks *via* the LDAR program or replacements of leaky components and pipelines. The emission reduction can also be done through a change of process and use of wasted emissions, for example, from flares and vents, which can have economical benefits [160]. It should be noted that emissions from past activities that are not anymore in the supply chain should also be taken into account, for example, from abandoned wells.

Atmospheric methane capture [161] requires the development of technologies to capture and remove methane molecules at the emitting sources or from ambient air. A low abundance of ambient methane is a challenge to achieving the goals of this technology. It should also be noted that this process should have fewer carbon footprints compared to the carbon equivalent captured and removed [162].

Atmospheric methane destruction can be done through geo-engineering, for example, the use of chlorine to destruct ambient methane molecules [163, 164]. This process can on paper reduce atmospheric methane abundance by increasing the oxidation process. However, care must be taken during the geoengineering activities, and further studies are required.

The transition from the use of fossil fuels to renewable energy has been started for some time already. However, long-term dependency on fossil fuels, wide availabilities and ease of extraction of fossil fuel resources, price fluctuations, challenges in financing the transition and societal impression are the barriers in the transition phase [165].

All these solutions have their advantages, disadvantages, ease and challenges, which, possibly all together with more weight on some, can have larger positive impacts with less negative results. With consideration of current situations, it can be recommended to first implement emission reduction efforts, secondly on energy transition and thirdly on developing technologies and techniques to capture, remove and destroy ambient methane.

Advancements in technologies, such as satellite-based measurements, can significantly accelerate the detection of large emitters, which may arise from various sources. These advancements enable the rapid notification of stakeholders, facilitating swift actions to reduce or eliminate emissions, such as through initiatives like the MARS program of UNEP-IMEO. Satellite technology generates vast amounts of data per satellite in real time, which can be efficiently processed and analyzed using AI-based approaches. Identifying emission sources is the first critical step toward achieving mitigation. However, the subsequent step—collaboration from operators to implement mitigation measures, such as repairing or replacing equipment at emission sources—remains underdeveloped [166].

Delays or failures in emission mitigation efforts can stem from various factors, including: (I) lack of mitigation policies in certain countries, (II) low prioritization of emission mitigation, (III) insufficient funding and/or human resources dedicated to mitigation, (IV) difficult access to emission points, (V) restricted access to technologies due to sanctions or financial constraints, (VI) reliance on outdated infrastructure, (VII) infrequent maintenance practices, (VIII) prolonged bureaucratic processes within the responsible operators, (IX) technical barriers arising from initial operational designs, such as the need for fundamental changes in operations to mitigate flare gases, (X) geopolitical conflicts that exacerbate the aforementioned challenges and (XI) political mindsets that delay or resist the implementation of environmental actions.

5. Methane emissions and geopolitical conditions and politicizing of environmental endeavors

At the current state of global warming (**Figure 1**), it can be seen that the global mean temperature is on the rise relative to past years, which has resulted in several extreme events in the climate change framework. This shows that extensive, rapid and in-depth actions are required to be implemented to avoid further increases in the context of sustainable developments.

While methane release at a point dilutes in the atmosphere and circles around the globe without considering the borders, geopolitical conflicts that are bound to the borders can lead to a halt and/or slowdown in addressing the emissions, global warming and climate change and even can result in increase in emissions [167, 168]. This shows that care and attentions are required to put forward in the context of international relationships, which is a necessity for everyone.

5.1 Exploitation of natural resources for domestic economic development

The management of natural resources, such as water, has been extensively studied through game theory, which demonstrates how systematic thinking or individual-benefit thinking among stakeholders can harm the collective benefits of a system [169]. In this context, the "system" is the atmosphere, and stakeholders' actions significantly impact the emissions contributing to the atmospheric abundance of GHGs. Although the exploitation of natural resources may provide individual benefits as suggested by game theory, it often exacerbates emissions and global warming, thereby damaging the system (**Figure 1**).

5.2 Geopolitical conflicts and politicization of environmental regulations

Two critical factors—geopolitical conflicts and the politicization of environmental regulations—can delay or obstruct emission mitigation efforts, or even exacerbate emissions. These factors can lead to: (I) deprioritization of environmental considerations, (II) changes in mitigation regulations, (III) negative impacts on mitigation dialogs, (IV) reductions in domestic budgets for sustainable development and research, (V) restricted access to mitigation technologies due to sanctions or financial constraints, particularly when technologies are labeled as dual purpose (environmental and military use), (VI) increased reliance on carbon-intensive energy production, (VII) demolition of infrastructure, requiring rebuilding that generates additional emissions, and (VIII) loss of human lives [170], which deprives societies of critical resources for advancing sustainable development.

5.2.1 Politicizing environmental endeavors and domestic emissions

For example, binding environmental policies to presidential cabinet changes can: (I) result in new regulations that may take time to develop, potentially increasing emissions during the transition period, and (II) lead to elevated emissions in subsequent years [171].

Similarly, domestic budget cuts can negatively impact sustainable development efforts, including emission mitigation [172]. Budget reductions can also affect higher education, which is fundamental to advancing our understanding of Earth systems and accelerating mitigation efforts [173]. Moreover, the construction of new carbon-intensive power plants can significantly increase a country's overall emissions [174].

5.2.2 Sanctions and domestic emissions

The impact of sanctions on domestic emissions in the energy sector can vary based on whether a country is a producer or importer of fossil fuels. Sanctions targeting fossil fuel industries, banking systems, or dual-purpose technologies can be categorized as follows:

a. Fossil fuel-producing countries for domestic use

Sanctions on these countries often result in a focus on exploiting fossil fuel resources, with environmental factors deprioritized. Over time, limited access to technologies and funds may degrade the energy supply chain's quality, increasing emissions.

b. Fossil fuel-producing countries for export

For these countries, a fossil fuel-based economy is heavily reliant on exports. Sanctions can reduce export revenues, leading to lower budget allocations for emission mitigation. Additionally, production levels may be reduced, or operations may continue at levels that generate excess emissions, as surplus fuel can no longer be exported.

c. Non-producing fossil fuel countries and importers of fossil fuels

The economic impact of sanctions on these countries depends on their reliance on fossil fuels for domestic energy. In severe cases, economic burdens may deprioritize environmental actions, even if such measures were previously in place. Some countries may shift to renewable energy sources based on domestic knowledge, which can have a positive impact.

d. Non-producing fossil fuel countries and non-importers of fossil fuels

For countries that rely on alternative energy sources, sanctions on fossil fuels may not directly increase emissions. However, broader impacts on the global economy could still indirectly affect their environmental efforts.

6. Recommendations

To facilitate emission mitigation, it is essential to address the gaps identified, which can be categorized into (I) knowledge gaps, (II) technology gaps and (III) policy gaps.

6.1 Knowledge gaps

Knowledge gaps can be addressed by implementing and continuing extensive measurement-based campaigns, particularly in under-studied countries and regions, and releasing the results in open-access publications. Conducting further snapshot measurements and continuous monitoring can be highly beneficial in filling these gaps. Cooperation from operators in providing transparent information and taking appropriate actions when notified about an emitter can also be invaluable.

6.2 Technology gaps

While current technologies are sufficient to detect and quantify large emitters, accessing these technologies remains a limiting factor for stakeholders due to financial constraints or sanctions. Access to these technologies should be facilitated through funding agencies, and relevant technology transfer opportunities should be provided. Further technological developments should focus on reducing costs and uncertainty ranges, making these developed technologies more accessible to all countries.

6.3 Policy gaps

Emission mitigation policies should be clearly depoliticized, emphasizing that emission reduction is vital for the overall benefit of the atmosphere. Policies should not be subject to government changes but rather progressively improved to reduce emissions consistently over time. The carbon footprint of geopolitical conflicts (violent or non-violent) should also be considered for accountable actions, as these conflicts can lead to increased GHG emissions and contribute to the rise in global mean temperatures (**Figure 1**).

7. Conclusion

Natural gas systems have been considered as a bridging solution toward the use of a sustainable energy supply. However, the current state of methane emissions from these systems can impose negative impacts on this bridging solution. To understand the situation, extensive, coordinated measurement-based campaigns are required to enlighten further. Methane regulations are of necessity to be scaled to several sectors and countries to minimize the emissions to the lowest. With the current state of technologies, it is possible to reduce emissions at the emitting points, which have environmental, financial and societal positive impacts. Further technological developments can possibly be an option to lower atmospheric methane abundance, which requires special cares due to the side effects. Mitigation actions are required to be implemented at the earliest time possible, and politicization of environmental endeavours specifically with regards to emission mitigation efforts and geo-political conditions slow down these efforts and/or increase emissions, which has negative impacts on everyone around the globe.

Acknowledgements

We thank Murès Zarea of the European Gas Research Group (GERG) and Jonathan Martinez of Bureau Veritas (BV) for their time reading the chapter and providing insightful comments and constructive feedbacks to enhance the work. This chapter is dedicated to the communities, individuals, and stakeholders who recognize the urgent need for emission mitigation efforts and are committed to paving the way toward decarbonization under real-world conditions, for the benefit of the shared global earth system, in particular atmosphere.

Conflict of interest

The authors declare no conflict of interest.

Financial support

This chapter receives financial support from Professorship of Environmental Sensing and Modeling at Technical University of Munich (TUM) and the Islamic Republic of Iran's vice presidency for Science and Technology through University of Tehran (UT). Jia Chen acknowledges funding from the European Union's Horizon 2020 project PAUL (Grant No. 101037319) and the European Research Council (ERC) Consolidator Grant CoSense4Climate (Grant No. 101089203).

Author details

Hossein Maazallahi[1*], Jia Chen[2] and Julianne M. Fernandez[3]

1 Department of Renewable Energies and Environment, College of Interdisciplinary Science and Technologies, University of Tehran, Tehran, Islamic Republic of Iran

2 Environmental Sensing and Modeling, Technical University of Munich, Munich, Germany

3 Scientific Aviation, Boulder, Colorado, United States of America

*Address all correspondence to: h.maazallahi@ut.ac.ir

IntechOpen

© 2025 The Author(s). Licensee IntechOpen. This chapter is distributed under the terms of the Creative Commons Attribution License (http://creativecommons.org/licenses/by/4.0), which permits unrestricted use, distribution, and reproduction in any medium, provided the original work is properly cited.

References

[1] Staniaszek Z, Griffiths PT, Folberth GA, O'Connor FM, Abraham NL, Archibald AT. The role of future anthropogenic methane emissions in air quality and climate. npj Climate and Atmospheric Science [Internet]. 2022;**5**(1):1-8. DOI: 10.1038/s41612-022-00247-5

[2] Rogelj J, Shindell D, Jiang K, Fifita S, Forster P, Ginzburg V, et al. Mitigation Pathways Compatible with 1.5°C in the Context of Sustainable Development. Geneva, Switzerland: IPCC/WMO; 2018

[3] Reshetnikov AI, Paramonova NN, Shashkov AA. An evaluation of historical methane emissions from the soviet gas industry. Journal of Geophysical Research: Atmospheres [Internet]. 2000;**105**(D3):3517-3529. DOI: 10.1029/1999JD900761

[4] Saunois M, Stavert RA, Poulter B, Bousquet P, Canadell JG, Jackson RB, et al. The global methane budget 2000-2017. Earth System Science Data. 2020;**12**(3):1561-1623. DOI: 10.5194/essd-12-1561-2020

[5] UNEP. Methane Action: Tackling a Warming Planet [Internet]. Paris, France: UNEP; 2021. Available from: https://www.unep.org/news-and-stories/speech/methane-action-tackling-warming-planet [Accessed: October 5, 2024]

[6] European Commission. Methane Emissions - European Commission [Internet]. Brussels, Belgium: European Commission; 2022. Available from: https://energy.ec.europa.eu/topics/carbon-management-and-fossil-fuels/methane-emissions_en [Accessed: October 5, 2024]

[7] United Nations. The Paris Agreement [Internet]. Paris, France: UNFCCC; 2015. Available from: https://unfccc.int/process-and-meetings/the-paris-agreement [Accessed: October 2, 2024]

[8] Munnings C, Krupnick A, Kruger J, George F, Energy S, Michels T, et al. Comparing Policies to Reduce Methane Emissions in the Natural Gas Sector. Washington, DC, United States of America: Resources for the Future; 2017

[9] European Commission. Deal on First-Ever EU Law to Curb Methane Emissions [Internet]. Brussels, Belgium: European Commission; 2023. Available from: https://ec.europa.eu/commission/presscorner/detail/en/ip_23_5776 [Accessed: October 5, 2024]

[10] European Commission. EU Methane Strategy [Internet]. Brussels, Belgium: European Commission; 2020. Available from: https://ec.europa.eu/commission/presscorner/api/files/attachment/866544/Methane_GreenDeal_EN.pdf [Accessed: October 5, 2024]

[11] European Union. EU Methane Regulation - EU - 2024/1787 - EN - EUR-Lex_REGULATION (EU) 2024/1787 of the European Parliament and of the Council [Internet]. Brussels, Belgium: European Union; 2024. Available from: https://eur-lex.europa.eu/legal-content/EN/TXT/?uri=OJ:L_202401787&pk_campaign=todays_OJ&pk_source=EURLEX&pk_medium=X&pk_keyword=energy_sector&pk_content=Regulation&pk_cid=EURLEX_todaysOJ [Accessed: October 5, 2024]

[12] European Commission. EC. Questions and Answers on REPowerEU: Joint European Action for More

Affordable, Secure and Sustainable Energy. Brussels, Belgium: European Commission; 2022. Available from: https://ec.europa.eu/commission/presscorner/detail/en/qanda_22_1512 [Accessed: October 5, 2024]

[13] Global Methane Pledge. Homepage. Global Methane Pledge [Internet]. Paris, France: Global Methane Pledge; 2021. Available from: https://www.globalmethanepledge.org/ [Accessed: October 5, 2024]

[14] Stechemesser A, Koch N, Mark E, Dilger E, Klösel P, Menicacci L, et al. Climate policies that achieved major emission reductions: Global evidence from two decades. Science [Internet]. 2024;**385**(6711):884-892. DOI: 10.1126/science.adl6547

[15] GERG. GERG. 2024. Gerg - The European Gas Research Group. Available from: https://www.gerg.eu/ [Accessed: October 6, 2024]

[16] Global Methane Hub. Global Methane Hub - Meet the Moment on Methane [Internet]. Washington, DC, United States of America: Global Methane Hub; 2024. Available from: https://www.globalmethanehub.org/ [Accessed: October 6, 2024]

[17] Global Methane Initiative. Global Methane Initiative [Internet]. Washington, DC, United States of America; 2024. Available from: https://www.globalmethane.org/ [Accessed: October 6, 2024]

[18] TADI. TADI: un centre d'essai d'envergure internationale. ETABLISSEMENT PAU LACQ [Internet]. 2024. Available from: https://cstjf-pau.totalenergies.fr/en/our-expertise/leveraging-digital-innovation/tadi-test-center-international-reach-lacq [Accessed: October 6, 2024]

[19] METEC. METEC. Colorado State University [Internet]. 2024. Available from: https://metec.colostate.edu/ [Accessed: October 6, 2024]

[20] IMEO. International Methane Emissions Observatory. UNEP - UN Environment Programme [Internet]. 2024. Available from: https://www.unep.org/topics/energy/methane/international-methane-emissions-observatory [Accessed: October 6, 2024]

[21] UNEP. UNEP. OGMP 2.0 – The Oil & Gas Methane Partnership 2.0. 2024. Available from: https://ogmpartnership.com/ [Accessed: October 6, 2024]

[22] Methane Mitigation Summit Series. Reducing Methane Emission [Internet]. London, United Kingdom: Methane Mitigation Summit Series; 2024. Available from: https://www.industrialdecarbonizationnetwork.com/events-methane-mitigation-summit-series [Accessed: October 6, 2024]

[23] Maazallahi H, Delre A, Scheutz C, Fredenslund AM, Schwietzke S, Denier Van Der Gon H, et al. Intercomparison of detection and quantification methods for methane emissions from the natural gas distribution network in Hamburg, Germany. Atmospheric Measurement Techniques. 2023;**16**(21):5051-5073. DOI: 10.5194/amt-16-5051-2023

[24] Commane R, Hallward-Driemeier A, Murray LT. Intercomparison of commercial analyzers for atmospheric ethane and methane observations. Atmospheric Measurement Techniques. 2023;**16**(5):1431-1441. DOI: 10.5194/amt-16-1431-2023

[25] Liu Y, Paris JD, Broquet G, Bescós Roy V, Fernandez TM, Andersen R, et al. Assessment of current methane emission quantification techniques for natural gas midstream applications.

Atmospheric Measurement Techniques. 2024;**17**(6):1633-1649. DOI: 10.5194/amt-17-1633-2024

[26] Improving Characterization of Anthropogenic Methane Emissions in the United States [Internet]. Improving Characterization of Anthropogenic Methane Emissions in the United States. Washington, DC, United States of America: National Academies Press; 2018. Available from: https://nap.nationalacademies.org/read/24987/chapter/5 [Accessed: December 10, 2024]

[27] GERG. GERG 'Technology Benchmark for Site Level Methane Emissions Quantification' – Phase II.B - Gerg [Internet]. Brussels, Belgium: GERG; 2024. Available from: https://www.gerg.eu/projects/methane-emissions/gerg-technology-benchmark-for-site-level-methane-emissions-quantification-phase-ii-b/ [Accessed: October 2, 2024]

[28] Russell M. The Nord Stream 2 Pipeline. Brussels, Belgium: European Parliamentary Research Service; 2021

[29] Schrider LA, Wise RL. Potential new sources of natural gas. Journal of Petroleum Technology [Internet]. 1980;**32**(04):703-716. DOI: 10.2118/7628-PA

[30] Brandt AR, Heath GA, Cooley D. Methane leaks from natural gas systems follow extreme distributions. Environmental Science & Technology [Internet]. 2016;**50**(22):12512-12520. DOI: 10.1021/acs.est.6b04303

[31] Mayfield EN, Robinson AL, Cohon JL. System-wide and superemitter policy options for the abatement of methane emissions from the U.S. natural gas system. Environmental Science & Technology [Internet]. 2017;**51**(9):4772-4780. DOI: 10.1021/acs.est.6b05052

[32] Howarth R, Shindell D, Santoro R, Ingraffea A, Phillips N, Townsend-Smal A. Methane Emissions from Natural Gas Systems Background Paper Prepared for the National Climate Assessment. Narrowsburg, NY, United States of America: Damascus Citizens for Sustainability; 2012

[33] Third National Communication to United Nations Framework Convention on Climate Change. Tehran, Islamic Republic of Iran. Department of Environment, National Climate Change Office; 2017. Available from: https://unfccc.int/sites/default/files/resource/Third%20National%20communication%20IRAN.pdf: [Accessed: February 8, 2025]

[34] Balcombe P, Anderson K, Speirs J, Brandon N, Hawkes A. The natural gas supply chain: The importance of methane and carbon dioxide emissions. ACS Sustainable Chemistry & Engineering [Internet]. 2017;**5**(1):3-20. DOI: 10.1021/acssuschemeng.6b00144

[35] MacKay K, Seymour SP, Li HZ, Zavala-Araiza D, Xie D. A comprehensive integration and synthesis of methane emissions from Canada's oil and gas value chain. Environmental Science & Technology [Internet]. 2024;**58**(32):14203-14213. DOI: 10.1021/acs.est.4c03651

[36] Schwietzke S, Pétron G, Conley S, Pickering C, Mielke-Maday I, Dlugokencky EJ, et al. Improved mechanistic understanding of natural gas methane emissions from spatially resolved aircraft measurements. Environmental Science & Technology [Internet]. 2017;**51**(12):7286-7294. DOI: 10.1021/acs.est.7b01810

[37] Miller SM, Wofsy SC, Michalak AM, Kort EA, Andrews AE, Biraud SC, et al. Anthropogenic emissions of methane

in the United States. Proceedings of the National Academy of Sciences of the United States of America [Internet]. 2013;**110**(50):20018-20022. DOI: 10.1073/pnas.1314392110

[38] Rutherford JS, Sherwin ED, Ravikumar AP, Heath GA, Englander J, Cooley D, et al. Closing the methane gap in US oil and natural gas production emissions inventories. Nature Communications [Internet]. 2021;**12**(1):1-12. DOI: 10.1038/s41467-021-25017-4

[39] Brandt AR, Heath GA, Kort EA, O'Sullivan F, Pétron G, Jordaan SM, et al. Methane leaks from North American natural gas systems. Science (1979) [Internet]. 2014;**343**(6172):733-735. DOI: 10.1126/science.1247045

[40] Allen DT. Methane emissions from natural gas production and use: Reconciling bottom-up and top-down measurements. Current Opinion in Chemical Engineering. 2014;**5**:78-83. DOI: 10.1016/j.coche.2014.05.004

[41] Omara M, Sullivan MR, Li X, Subramian R, Robinson AL, Presto AA. Methane emissions from conventional and unconventional natural gas production sites in the Marcellus Shale Basin. Environmental Science & Technology [Internet]. 2016;**50**(4):2099-2107. DOI: 10.1021/acs.est.5b05503

[42] Barkley ZR, Lauvaux T, Davis KJ, Deng A, Miles NL, Richardson SJ, et al. Quantifying methane emissions from natural gas production in North-Eastern Pennsylvania. Atmospheric Chemistry and Physics. 2017;**17**(22):13941-13966. DOI: 10.5194/acp-17-13941-2017

[43] Karion A, Sweeney C, Pétron G, Frost G, Michael Hardesty R, Kofler J, et al. Methane emissions estimate from airborne measurements over a western United States natural gas field. Geophysical Research Letters [Internet]. 2013;**40**(16):4393-4397. DOI: 10.1002/grl.50811

[44] Howarth RW, Robert Howarth CW. A bridge to nowhere: Methane emissions and the greenhouse gas footprint of natural gas. Energy Science & Engineering [Internet]. 2014;**2**(2):47-60. DOI: 10.1002/ese3.35

[45] Maazallahi H. Switching to the "Golden Age of Natural Gas" with a Focus on Shale Gas Exploitation: A Possible Bridge to Mitigate Climate Change? [Internet], Lund University GEM thesis Series. Lund: Lund University. 2015. Available from: http://lup.lub.lu.se/student-papers/record/7989502 [Accessed: October 6, 2024]

[46] Harrison MR, Shires TM, Wessels JK, Cowgill RM. Methane Emissions from the Natural Gas Industry Project Summary. Washington, DC, United States of America: United States Environmental Protection Agency; 1997. Available from: https://p2infohouse.org/ref/07/06348.pdf [Accessed: October 2, 2024]

[47] Beck LL. A global methane emissions program for landfills, coal mines, and natural gas systems. Chemosphere. 1993;**26**(1-4):447-452. DOI: 10.1016/0045-6535(93)90437-A

[48] Kirchgessner DA, Lott RA, Cowgill RM, Harrison MR, Shires TM. Estimate of methane emissions from the U.S. natural gas industry. Chemosphere. 1997;**35**(6):1365-1390. DOI: 10.1016/S0045-6535(97)00236-1

[49] Dedikov JV, Akopova GS, Gladkaja NG, Piotrovskij AS, Markellov VA, Salichov SS, et al. Estimating methane releases from

natural gas production and transmission in Russia. Atmospheric Environment. 1999;**33**(20):3291-3299. DOI: 10.1016/S1352-2310(98)00388-4

[50] Lelieveld J, Lechtenböhmer S, Assonov SS, Brenninkmeijer CAM, Dienst C, Fischedick M, et al. Low methane leakage from gas pipelines. Nature [Internet]. 2005;**434**(7035):841-842. DOI: 10.1038/434841a

[51] Mitchell C, Sweet J, Jackson T. A study of leakage from the UK natural gas distribution system. Energy Policy. 1990;**18**(9):809-818. DOI: 10.1016/0301-4215(90)90060-H

[52] Wunch D, Wennberg PO, Toon GC, Keppel-Aleks G, Yavin YG. Emissions of greenhouse gases from a North American megacity. Geophysical Research Letters [Internet]. 2009;**36**(15):L15810. DOI: 10.1029/2009GL039825

[53] MEMO2. MEthane goes MObile, MEasurements and MOdelling. European Union's Horizon 2020, Marie Skłodowska-Curie Grant [Internet]. 2017. Available from: https://h2020-memo2.eu [Accessed: March 7, 2025]

[54] Fletcher SEM, Schaefer H. Rising methane: A new climate challenge. Science (1979) [Internet]. 2019;**364**(6444):932-933. DOI: 10.1126/science.aax1828

[55] Leak Detection and Repair Compliance Assistance Guidance Best Practices Guide [Internet]. 2014. Available from: https://www.epa.gov/sites/default/files/2014-02/documents/ldarguide.pdf [Accessed: October 2, 2024]

[56] Cheadle LC, Tran T, Nyarady JF, Lozo C. Leak detection and repair data from California's oil and gas methane regulation show decrease in leaks over two years. Environmental Challenges. 2022;**8**:100563. DOI: 10.1016/j.envc.2022.100563

[57] Schuit BJ, Maasakkers JD, Bijl P, Mahapatra G, Van Den Berg AW, Pandey S, et al. Automated detection and monitoring of methane super-emitters using satellite data. Atmospheric Chemistry and Physics. 2023;**23**(16):9071-9098. DOI: 10.5194/acp-23-9071-2023

[58] Vaughan A, Mateo-García G, Gómez-Chova L, Ruzička V, Guanter L, Irakulis-Loitxate I. CH4Net: A deep learning model for monitoring methane super-emitters with Sentinel-2 imagery. Atmospheric Measurement Techniques. 2024;**17**(9):2583-2593. DOI: 10.5194/amt-17-2583-2024

[59] Bretschneider L, Bollmann S, Houssin-Agbomson D, Shaw J, Howes N, Nguyen L, et al. Concepts for drone based pipeline leak detection. Frontiers in Robotics and AI. 2024;**11**:1426206. DOI: 10.3389/frobt.2024.1426206

[60] Roy DP, Wulder MA, Loveland TR, Woodcock CE, Allen RG, Anderson MC, et al. Landsat-8: Science and product vision for terrestrial global change research. Remote Sensing of Environment. 2014;**145**:154-172. DOI: 10.1016/j.rse.2014.02.001

[61] Drusch M, Del Bello U, Carlier S, Colin O, Fernandez V, Gascon F, et al. Sentinel-2: ESA's optical high-resolution mission for GMES operational services. Remote Sensing of Environment. 2012;**120**:25-36. DOI: 10.1016/j.rse.2011.11.026

[62] Veefkind JP, Aben I, McMullan K, Förster H, de Vries J, Otter G, et al. TROPOMI on the ESA Sentinel-5 precursor: A GMES mission for global observations of the atmospheric

composition for climate, air quality and ozone layer applications. Remote Sensing of Environment. 2012;**120**:70-83. DOI: 10.1016/j.rse.2011.09.027

[63] Lorente A, Borsdorff T, Butz A, Hasekamp O, Aan De Brugh J, Schneider A, et al. Methane retrieved from TROPOMI: Improvement of the data product and validation of the first 2 years of measurements. Atmospheric Measurement Techniques. 2021;**14**(1):665-684. DOI: 10.5194/amt-14-665-2021

[64] Galeazzi C, Sacchetti A, Cisbani A, Babini G. The prisma program. International Geoscience and Remote Sensing Symposium (IGARSS). 2008;4(1). DOI: 10.1109/IGARSS.2008.4779667

[65] Jervis D, McKeever J, Durak BOA, Sloan JJ, Gains D, Varon DJ, et al. The GHGSat-D imaging spectrometer. Atmospheric Measurement Techniques. 2021;**14**(3):2127-2140. DOI: 10.5194/amt-14-2127-2021

[66] MethaneSat. MethaneSat [Internet]. 2024. Available from: www.methanesat.org [Accessed: October 6, 2024]

[67] Carbon Mapper. Carbon Mapper [Internet]. 2024. Available from: https://carbonmapper.org/ [Accessed: October 6, 2024]

[68] Fredenslund AM, Scheutz C, Kjeldsen P. Tracer method to measure landfill gas emissions from leachate collection systems. Waste Management. 2010;**30**(11):2146-2152. DOI: 10.1016/j.wasman.2010.03.013

[69] Fredenslund AM, Rees-White TC, Beaven RP, Delre A, Finlayson A, Helmore J, et al. Validation and error assessment of the mobile tracer gas dispersion method for measurement of fugitive emissions from area sources. Waste Management. 2019;**83**:68-78. DOI: 10.1016/j.wasman.2018.10.036

[70] Awasthi S, Khare M, Gargava P. General plume dispersion model (GPDM) for point source emission. Environmental Modeling and Assessment. 2006;**11**:267-276. DOI: 10.1007/s10666-006-9041-y

[71] Korbeń P, Jagoda P, Maazallahi H, Kammerer J, Nęcki JM, Wietzel JB, et al. Quantification of methane emission rate from oil and gas wells in Romania using ground-based measurement techniques. Elementa [Internet]. 2022;**10**(1):00070. DOI: 10.1525/elementa.2022.00070

[72] Cho Y, Ulrich BA, Zimmerle DJ, Smits KM. Estimating natural gas emissions from underground pipelines using surface concentration measurements☆. Environmental Pollution. 2020;**267**:115514. DOI: 10.1016/j.envpol.2020.115514

[73] Weller ZD, Yang DK, Von Fischer JC. An open source algorithm to detect natural gas leaks from mobile methane survey data. PLoS One [Internet]. 2019;**14**(2):e0212287. DOI: 10.1371/journal.pone.0212287

[74] Lamb BK, Mcmanus JB, Shorter JH, Kolb CE, Mosher B, Harriss RC, et al. Development of atmospheric tracer methods to measure methane emissions from natural gas facilities and urban areas. Environmental Science & Technology [Internet]. 1995;**29**(6):1468-1479. DOI: 10.1021/es00006a007

[75] Innocenti F, Robinson R, Gardiner T, Finlayson A, Connor A. Differential absorption Lidar (DIAL) measurements of landfill methane emissions. Remote Sensing [Internet]. 2017;**9**(9):953. DOI: 10.3390/rs9090953

[76] GERG. Quantifying Underground Leakages from (Gas) Pipelines [Internet]. Brussels, Belgium: GERG; 2020. Available from: https://www.gerg.eu/projects/methane-emissions/quantifying-underground-leakages-from-gas-pipelines/ [Accessed: October 5, 2024]

[77] Zirnig W, Ulbricht M. Innovative technologies improve environmental protection-detection of gas leaks by helicopter-borne infrared laser system. In: IGU - WGC [Internet]. Tokyo: IGU; 2003. Available from: http://members.igu.org/html/wgc2003/WGC_pdffiles/10154_1044541118_18323_1.pdf [Accessed: October 5, 2024]

[78] Yacovitch TI, Daube C, Herndon SC. Methane emissions from offshore oil and gas platforms in the Gulf of Mexico. Environmental Science & Technology [Internet]. 2020;**54**(6):3530-3538. DOI: 10.1021/acs.est.9b07148

[79] Nara H, Tanimoto H, Tohjima Y, Mukai H, Nojiri Y, Machida T. Emissions of methane from offshore oil and gas platforms in Southeast Asia. Scientific Reports [Internet]. 2014;**4**(1):1-6. DOI: 10.1038/srep06503

[80] Hensen A, Velzeboer I, Frumau KFA, Van den Bulk WCM, Van Dinther D. Methane Emission Measurements of Offshore Oil and Gas Platforms [Internet]. Petten: TNO; 2019. Available from: https://publications.tno.nl/publication/34634946/XYcOP9/TNO-2019-R10895.pdf [Accessed: October 29, 2024]

[81] de Groot TR, Menoud M, van Bleijswijk J, van Leeuwen SM, van der Molen J, Hernando-Morales V, et al. Tidal and seasonal influence on cold seep activity and methanotroph efficiency in the North Sea. Communications Earth & Environment [Internet]. 2024;**5**(1):1-13. DOI: 10.1038/s43247-024-01483-8

[82] Workman E, Fisher RE, France JL, Linse K, Yang M, Bell T, et al. Methane emissions from seabed to atmosphere in polar oceans revealed by direct methane flux measurements. Journal of Geophysical Research: Atmospheres [Internet]. 2024;**129**(14):e2023JD040632. DOI: 10.1029/2023JD040632

[83] Vinković K, Andersen T, de Vries M, Kers B, van Heuven S, Peters W, et al. Evaluating the use of an unmanned aerial vehicle (UAV)-based active AirCore system to quantify methane emissions from dairy cows. Science of The Total Environment. 2022;**831**:154898. DOI: 10.1016/j.scitotenv.2022.154898

[84] Mohammadloo TH, Jones MJ, Van de Kerkhof B, Dawson K, Smith BJ, Conley S, et al. Quantitative estimate of sources of uncertainty in drone-based methane emission measurements. EGUsphere [preprint]. 2024:egusphere-2024-1175. DOI: 10.5194/egusphere-2024-1175

[85] Tassielli G, Cananà L, Spalatro M. Detection of methane leaks via a drone-based system for sustainable landfills and oil and gas facilities: Effect of different variables on the background-noise measurement. Sustainability [Internet]. 2024;**16**(17):7748. DOI: 10.3390/su16177748

[86] Shi T, Han Z, Han G, Ma X, Chen H, Andersen T, et al. Retrieving CH4-emission rates from coal mine ventilation shafts using UAV-based AirCore observations and the genetic algorithm-interior point penalty function (GA-IPPF) model. Atmospheric Chemistry and Physics. 2022;**22**(20):13881-13896. DOI: 10.5194/acp-22-13881-2022

[87] Förster E, Huntrieser H, Lichtenstern M, Pätzold F, Bretschneider L, Lampert A, et al. A novel helicopter-borne application for quantifying methane emissions from industrial activities: Results from measurements of coal mine ventilation shafts in Poland. In: EGU General Assembly. 2024:Article No. EGU24-15785. DOI: 10.5194/egusphere-egu24-15785

[88] Pekney NJ, Diehl JR, Ruehl D, Sams J, Veloski G, Patel A, et al. Measurement of methane emissions from abandoned oil and gas wells in Hillman State Park, Pennsylvania. Carbon Management [Internet]. 2018;**9**(2):165-175. DOI: 10.1080/17583004.2018.1443642

[89] Foulds A, Allen G, Shaw JT, Bateson P, Barker PA, Huang L, et al. Quantification and assessment of methane emissions from offshore oil and gas facilities on the Norwegian continental shelf. Atmospheric Chemistry and Physics. 2022;**22**(7):4303-4322. DOI: 10.5194/acp-22-4303-2022

[90] Bradley ES, Leifer I, Roberts DA, Dennison PE, Washburn L. Detection of marine methane emissions with AVIRIS band ratios. Geophysical Research Letters [Internet]. 2011;**38**(10):L10702. DOI: 10.1029/2011GL046729

[91] Thorpe AK, O'Handley C, Emmitt GD, DeCola PL, Hopkins FM, Yadav V, et al. Improved methane emission estimates using AVIRIS-NG and an airborne Doppler wind Lidar. Remote Sensing of Environment. 2021;**266**:112681. DOI: 10.1016/j.rse.2021.112681

[92] Wong G, Wang H, Park M, Park J, Ahn JY, Sung M, et al. Optimizing an airborne mass-balance methodology for accurate emission rate quantification of industrial facilities: A case study of industrial facilities in South Korea. Science of The Total Environment. 2024;**912**:169204. DOI: 10.1016/j.scitotenv.2023.169204

[93] Maazallahi H, Stavropoulou F, Sutanto SJ, Steiner M, Brunner D, Mertens M, et al. Airborne *in-situ* quantification of methane emissions from oil and gas production in Romania. Atmospheric Chemistry and Physics. 2025;**25**:1497-1511. DOI: 10.5194/acp-25-1497-2025

[94] Nesser H, Jacob DJ, Maasakkers JD, Lorente A, Chen Z, Lu X, et al. High-resolution US methane emissions inferred from an inversion of 2019 TROPOMI satellite data: Contributions from individual states, urban areas, and landfills. Atmospheric Chemistry and Physics. 2024;**24**(8):5069-5091. DOI: 10.5194/acp-24-5069-2024

[95] Hemati M, Mahdianpari M, Nassar R, Shiri H, Mohammadimanesh F. Urban methane emission monitoring across North America using TROPOMI data: An analytical inversion approach. Scientific Reports [Internet]. 2024;**14**(1):1-16. DOI: 10.1038/s41598-024-58995-8

[96] Guanter L, Roger J, Sharma S, Valverde A, Irakulis-Loitxate I, Gorroño J, et al. Multisatellite data depicts a record-breaking methane leak from a well blowout. Environmental Science & Technology Letters [Internet]. 2024;**11**(8):825-830. DOI: 10.1021/acs.estlett.4c00399

[97] MacLean JPW, Girard M, Jervis D, Marshall D, McKeever J, Ramier A, et al. Offshore methane detection and quantification from space using sun glint measurements with the GHGSat constellation. Atmospheric Measurement Techniques. 2024;**17**(2):863-874. DOI: 10.5194/amt-17-863-2024

[98] Dogniaux M, Maasakkers JD, Varon DJ, Aben I. Report on Landsat

8 and sentinel-2B observations of the Nord stream 2 pipeline methane leak. Atmospheric Measurement Techniques. 2024;**17**(9):2777-2787. DOI: 10.5194/amt-17-2777-2024

[99] Irakulis-Loitxate I, Gorroño J, Zavala-Araiza D, Guanter L. Satellites detect a methane ultra-emission event from an offshore platform in the Gulf of Mexico. Environmental Science & Technology Letters [Internet]. 2022;**9**(6):520-525. DOI: 10.1021/acs.estlett.2c00225

[100] Sherwin ED, Rutherford JS, Chen Y, Aminfard S, Kort EA, Jackson RB, et al. Single-blind validation of space-based point-source detection and quantification of onshore methane emissions. Scientific Reports [Internet]. 2023;**13**(1):1-10. DOI: 10.1038/s41598-023-30761-2

[101] UNEP-IMEO. IMEO Methane Data [Internet]. Paris, France: UNEP-IMEO; 2024. Available from: https://methanedata.unep.org/welcome [Accessed: October 6, 2024]

[102] ESA. GHGSat - Earth Online [Internet]. Paris, France: ESA; 2024. Available from: https://earth.esa.int/eogateway/missions/ghgsat [Accessed: October 6, 2024]

[103] Rouet-Leduc B, Hulbert C. Automatic detection of methane emissions in multispectral satellite imagery using a vision transformer. Nature Communications [Internet]. 2024;**15**(1):1-9. DOI: 10.1038/s41467-024-47754-y

[104] Chen J, Viatte C, Hedelius JK, Jones T, Franklin JE, Parker H, et al. Differential column measurements using compact solar-tracking spectrometers. Atmospheric Chemistry and Physics. 2016;**16**(13):8479-8498. DOI: 10.5194/acp-16-8479-2016

[105] Toja-Silva F, Chen J, Hachinger S, Hase F. CFD simulation of CO_2 dispersion from urban thermal power plant: Analysis of turbulent Schmidt number and comparison with Gaussian plume model and measurements. Journal of Wind Engineering and Industrial Aerodynamics. 2017;**169**:177-193. DOI: 10.1016/j.jweia.2017.07.015

[106] Luther A, Kostinek J, Kleinschek R, Defratyka S, Stanisavljević M, Forstmaier A, et al. Observational constraints on methane emissions from polish coal mines using a ground-based remote sensing network. Atmospheric Chemistry and Physics. 2022;**22**(9):5859-5876. DOI: 10.5194/acp-22-5859-2022

[107] Forstmaier A, Chen J, Dietrich F, Bettinelli J, Maazallahi H, Schneider C, et al. Quantification of methane emissions in Hamburg using a network of FTIR spectrometers and an inverse modeling approach. Atmospheric Chemistry and Physics. 2023;**23**(12):6897-6922. DOI: 10.5194/acp-23-6897-2023

[108] Hase F, Frey M, Blumenstock T, Groß J, Kiel M, Kohlhepp R, et al. Application of portable FTIR spectrometers for detecting greenhouse gas emissions of the major city Berlin. Atmospheric Measurement Techniques. 2015;**8**(7):3059-3068. DOI: 10.5194/amt-8-3059-2015

[109] Dietrich F, Chen J, Voggenreiter B, Aigner P, Nachtigall N. MUCCnet: Munich urban carbon column network. Atmospheric Measurement Techniques. 2021;**14**(2):1111-1126. DOI: 10.5194/amt-14-1111-2021

[110] Franco B, Mahieu E, Emmons LK, Tzompa-Sosa ZA, Fischer EV, Sudo K, et al. Evaluating ethane and methane emissions associated with the development of oil and natural

gas extraction in North America. Environmental Research Letters [Internet]. 2016;**11**(4):044010. DOI: 10.1088/1748-9326/11/4/044010

[111] Chen J, Dietrich F, Maazallahi H, Forstmaier A, Winkler D, Hofmann MEG, et al. Methane emissions from the Munich Oktoberfest. Atmospheric Chemistry and Physics. 2020;**20**(6):3683-3696. DOI: 10.5194/acp-20-3683-2020

[112] Dietrich F, Chen J, Shekhar A, Lober S, Krämer K, Leggett G, et al. Climate impact comparison of electric and gas-powered end-user appliances. Earths Future [Internet]. 2023;**11**(2):e2022EF002877. DOI: 10.1029/2022EF002877

[113] Rißmann M, Chen J, Osterman G, Zhao X, Dietrich F, Makowski M, et al. Comparison of OCO-2 target observations to MUCCnet - is it possible to capture urban XCO2gradients from space? Atmospheric Measurement Techniques. 2022;**15**(22):6605-6623. DOI: 10.5194/amt-15-6605-2022

[114] Schoell M. The hydrogen and carbon isotopic composition of methane from natural gases of various origins. Geochimica et Cosmochimica Acta. 1980;**44**(5):649-661. DOI: 10.1016/0016-7037(80)90155-6

[115] Menoud M, Van Der Veen C, Lowry D, Fernandez JM, Bakkaloglu S, France JL, et al. New contributions of measurements in Europe to the global inventory of the stable isotopic composition of methane. Earth System Science Data. 2022;**14**(9):4365-4386. DOI: 10.5194/essd-14-4365-2022

[116] Maazallahi H, Fernandez JM, Menoud M, Zavala-Araiza D, Weller ZD, Schwietzke S, et al. Methane mapping, emission quantification, and attribution in two European cities: Utrecht (NL) and Hamburg (DE). Atmospheric Chemistry and Physics. 2020;**20**(23):14717-14740. DOI: 10.5194/acp-20-14717-2020

[117] Schiferl LD, Cao C, Dalton B, Hallward-Driemeier A, Toledo-Crow R, Commane R. Multi-year observations of variable incomplete combustion in the New York megacity. Atmospheric Chemistry and Physics [Internet]. 2024;**24**(17):10129-10142. DOI: 10.5194/acp-24-10129-2024

[118] Fiehn A, Eckl M, Kostinek J, Gałkowski M, Gerbig C, Rothe M, et al. Source apportionment of methane emissions from the upper Silesian Coal basin using isotopic signatures. Atmospheric Chemistry and Physics. 2023;**23**(24):15749-15765. DOI: 10.5194/acp-23-15749-2023

[119] Ramachandran N, Irvin J, Omara M, Gautam R, Meisenhelder K, Rostami E, et al. Deep learning for detecting and characterizing oil and gas well pads in satellite imagery. Nature Communications [Internet]. 2024;**15**(1):1-14. DOI: 10.1038/s41467-024-50334-9

[120] Saneiyan S, Mansourian D. Locating undocumented orphaned oil and gas wells with smartphones. Journal of Applied Geophysics. 2023;**219**:105224. DOI: 10.1016/j.jappgeo.2023.105224

[121] Karion A, Link MF, Robertson R, Boyle T, Poppendieck D. Methodology and uncertainty estimation for measurements of methane leakage in a manufactured house. Atmospheric Measurement Techniques [Internet]. 2024;**17**:7065-7075. DOI: 10.5194/amt-17-7065-2024

[122] Riddick SN, Mbua M, Riddick JC, Houlihan C, Hodshire AL, Zimmerle DJ. Uncertainty quantification of methods

used to measure methane emissions of 1 g CH4 h−1. Sensors [Internet]. 2023;**23**(22):9246. DOI: 10.3390/s23229246

[123] Jahan I, Mehana M, Matheou G, Viswanathan H. Deep learning-based quantifications of methane emissions with field applications. International Journal of Applied Earth Observation and Geoinformation. 2024;**132**:104018. DOI: 10.1016/j.jag.2024.104018

[124] Morales R, Ravelid J, Vinkovic K, Korbeń P, Tuzson B, Emmenegger L, et al. Controlled-release experiment to investigate uncertainties in UAV-based emission quantification for methane point sources. Atmospheric Measurement Techniques. 2022;**15**(7):2177-2198. DOI: 10.5194/amt-15-2177-2022

[125] Wang JL, Daniels WS, Hammerling DM, Harrison M, Burmaster K, George FC, et al. Multiscale methane measurements at oil and gas facilities reveal necessary frameworks for improved emissions accounting. Environmental Science & Technology [Internet]. 2022;**56**(20):14743-14752. DOI: 10.1021/acs.est.2c06211

[126] Schissel C, Allen DT. Impact of the high-emission event duration and sampling frequency on the uncertainty in emission estimates. Environmental Science & Technology Letters [Internet]. 2022;**9**(12):1063-1067. DOI: 10.1021/acs.estlett.2c00731

[127] Schissel C, Allen D, Dieter H. Methods for spatial extrapolation of methane measurements in constructing regional estimates from sample populations. Environmental Science & Technology [Internet]. 2024;**58**(6):2739-2749. DOI: 10.1021/acs.est.3c08185

[128] Vaughn TL, Bell CS, Pickering CK, Schwietzke S, Heath GA, Pétron G, et al. Temporal variability largely explains top-down/bottom-up difference in methane emission estimates from a natural gas production region. Proceedings of the National Academy of Sciences of the United States of America [Internet]. 2018;**115**(46):11712-11717. DOI: 10.1073/pnas.1805687115

[129] Vollrath C, Hugenholtz CH, Barchyn TE. Onshore methane emissions measurements from the oil and gas industry: A scoping review. Environmental Research Communications [Internet]. 2024;**6**(3):032001. DOI: 10.1088/2515-7620/ad3129

[130] Stavropoulou F, Vinković K, Kers B, De Vries M, Van Heuven S, Korbeń P, et al. High potential for CH4 emission mitigation from oil infrastructure in one of EU's major production regions. Atmospheric Chemistry and Physics. 2023;**23**(18):10399-10412. DOI: 10.5194/acp-23-10399-2023

[131] Omara M, Zimmerman N, Sullivan MR, Li X, Ellis A, Cesa R, et al. Methane emissions from natural gas production sites in the United States: Data synthesis and National Estimate. Environmental Science & Technology [Internet]. 2018;**52**(21):12915-12925. DOI: 10.1021/acs.est.8b03535

[132] Shojaei SM, Vahabpour A, Saifoddin AA, Ghasempour R. Estimation of greenhouse gas emissions from Iran's gas flaring by using satellite data and combustion equations. Integrated Environmental Assessment and Management [Internet]. 2023;**19**(3):735-748. DOI: 10.1002/ieam.4684

[133] Roscioli JR, Yacovitch TI, Floerchinger C, Mitchell AL, Tkacik DS, Subramanian R, et al. Measurements of methane emissions from natural

gas gathering facilities and processing plants: Measurement methods. Atmospheric Measurement Techniques. 2015;**8**(5):2017-2035. DOI: 10.5194/amt-8-2017-2015

[134] Mitchell AL, Tkacik DS, Roscioli JR, Herndon SC, Yacovitch TI, Martinez DM, et al. Measurements of methane emissions from natural gas gathering facilities and processing plants: Measurement results. Environmental Science & Technology [Internet]. 2015;**49**(5):3219-3227. DOI: 10.1021/es5052809

[135] Marchese AJ, Vaughn TL, Zimmerle DJ, Martinez DM, Williams LL, Robinson AL, et al. Methane emissions from United States natural gas gathering and processing. Environmental Science & Technology [Internet]. 2015;**49**(17):10718-10727. DOI: 10.1021/acs.est.5b02275

[136] Zimmerle DJ, Williams LL, Vaughn TL, Quinn C, Subramanian R, Duggan GP, et al. Methane emissions from the natural gas transmission and storage system in the United States. Environmental Science & Technology [Internet]. 2015;**49**(15):9374-9383. DOI: 10.1021/acs.est.5b01669

[137] Subramanian R, Williams LL, Vaughn TL, Zimmerle D, Roscioli JR, Herndon SC, et al. Methane emissions from natural gas compressor stations in the transmission and storage sector: Measurements and comparisons with the EPA greenhouse gas reporting program protocol. Environmental Science & Technology [Internet]. 2015;**49**(5):3252-3261. DOI: 10.1021/es5060258

[138] Lamb BK, Edburg SL, Ferrara TW, Howard T, Harrison MR, Kolb CE, et al. Direct measurements show decreasing methane emissions from natural gas local distribution systems in the United States. Environmental Science & Technology [Internet]. 2015;**49**(8):5161-5169. DOI: 10.1021/es505116p

[139] Maazallahi H. Detection, Attribution and Quantification of Methane Emissions Using Mobile Measurement Techniques in European Cities [Internet]. Utrecht: Utrecht University; 2022. DOI: 10.33540/1457

[140] Afshoun HR, Jabbari SM, Jalaei SH. A regional study of the fugitive methane emissions from the gas distribution system in Iran by direct measurement. Journal of Natural Gas Science and Engineering. 2022;**105**:104716. DOI: 10.1016/j.jngse.2022.104716

[141] Von Fischer JC, Cooley D, Chamberlain S, Gaylord A, Griebenow CJ, Hamburg SP, et al. Rapid, vehicle-based identification of location and magnitude of urban natural gas pipeline leaks. Environmental Science & Technology [Internet]. 2017;**51**(7):4091-4099. DOI: 10.1021/acs.est.6b06095

[142] McKain K, Down A, Raciti SM, Budney J, Hutyra LR, Floerchinger C, et al. Methane emissions from natural gas infrastructure and use in the urban region of Boston, Massachusetts. Proceedings of the National Academy of Sciences of the United States of America [Internet]. 2015;**112**(7):1941-1946. DOI: 10.1073/pnas.1416261112

[143] Weller ZD, Hamburg SP, Von Fischer JC. A national estimate of methane leakage from pipeline mains in natural gas local distribution systems. Environmental Science & Technology [Internet]. 2020;**54**(14):8958-8967. DOI: 10.1021/acs.est.0c00437

[144] Merrin Z, Francisco PW. Unburned methane emissions from residential natural gas appliances. Environmental Science & Technology [Internet]. 2019;**53**(9):5473-5482. DOI: 10.1021/acs.est.8b05323

[145] Lebel ED, Finnegan CJ, Ouyang Z, Jackson RB. Methane and NOx emissions from natural gas stoves, cooktops, and ovens in residential homes. Environmental Science & Technology [Internet]. 2022;**56**(4):2529-2539. DOI: 10.1021/acs.est.1c04707

[146] Fischer ML, Chan WR, Delp W, Jeong S, Rapp V, Zhu Z. An estimate of natural gas methane emissions from California homes. Environmental Science & Technology [Internet]. 2018;**52**(17):10205-10213. DOI: 10.1021/acs.est.8b03217

[147] Wietzel JB, Schmidt M. Methane emission mapping and quantification in two medium-sized cities in Germany: Heidelberg and Schwetzingen. Atmospheric Environment: X. 2023;**20**:100228. DOI: 10.1016/j.aeaoa.2023.100228

[148] Joo J, Jeong S, Shin J, Chang DY. Missing methane emissions from urban sewer networks. Environmental Pollution. 2024;**342**:123101. DOI: 10.1016/j.envpol.2023.123101

[149] Fernandez JM, Maazallahi H, France JL, Menoud M, Corbu M, Ardelean M, et al. Street-level methane emissions of Bucharest, Romania and the dominance of urban wastewater. Atmospheric Environment: X. 2022;**13**:100153. DOI: 10.1016/j.aeaoa.2022.100153

[150] Defratyka SM, Paris JD, Yver-Kwok C, Fernandez JM, Korben P, Bousquet P. Mapping urban methane sources in Paris, France. Environmental Science & Technology [Internet]. 2021;**55**(13):8583-8591. DOI: 10.1021/acs.est.1c00859

[151] Vogel F, Ars S, Wunch D, Lavoie J, Gillespie L, Maazallahi H, et al. Ground-based Mobile measurements to track urban methane emissions from natural gas in 12 cities across eight countries. Environmental Science & Technology [Internet]. 2024;**58**(5):2271-2281. DOI: 10.1021/acs.est.3c03160

[152] Omara M, Zavala-Araiza D, Lyon DR, Hmiel B, Roberts KA, Hamburg SP. Methane emissions from US low production oil and natural gas well sites. Nature Communications [Internet]. 2022;**13**(1):1-10. DOI: 10.1038/s41467-022-29709-3

[153] Williams JP, Regehr A, Kang M. Methane emissions from abandoned oil and gas Wells in Canada and the United States. Environmental Science & Technology [Internet]. 2021;**55**(1):563-570. DOI: 10.1021/acs.est.0c04265

[154] Abrahamsson K, Damm E, Björk G, Bunse C, Sellmaier S, Broström G, et al. Methane plume detection after the 2022 Nord stream pipeline explosion in the Baltic Sea. Scientific Reports [Internet]. 2024;**14**(1):1-6. DOI: 10.1038/s41598-024-63449-2

[155] Kouznetsov R, Hänninen R, Uppstu A, Kadantsev E, Fatahi Y, Prank M, et al. A bottom-up emission estimate for the 2022 Nord stream gas leak: Derivation, simulations, and evaluation. Atmospheric Chemistry and Physics. 2024;**24**(8):4675-4691. DOI: 10.5194/acp-24-4675-2024

[156] Dissanayake AL, Gros J, Drews HJ, Nielsen JW, Drews A. Fate of methane from the Nord stream pipeline leaks. Environmental Science & Technology Letters [Internet]. 2023;**10**(10):903-908. DOI: 10.1021/acs.estlett.3c00493

[157] Poursanidis K, Sharanik J, Hadjistassou C. World's largest natural gas leak from Nord stream pipeline estimated at 478,000 tonnes. iScience. 2024;**27**(1):108772. DOI: 10.1016/j.isci.2023.108772

[158] UNEP. Impact of the Nord Stream Gas Leak on Methane Emissions. Paris, France: UNEP; 2023. Available from: https://www.unep.org/technical-highlight/impact-nord-stream-gas-leak-methane-emissions [Accessed: October 6, 2024]

[159] Harris SJ, Schwietzke S, France JL, Velandia Salinas N, Meixus Fernandez T, Randles C, et al. Methane emissions from the Nord Stream subsea pipeline leaks. Nature [Internet]. 2025;**637**(8048):1124-1130. DOI: 10.1038/s41586-024-08396-8

[160] Johnson MR, Coderre AR. Opportunities for CO_2 equivalent emissions reductions via flare and vent mitigation: A case study for Alberta, Canada. International Journal of Greenhouse Gas Control. 2012;**8**:121-131. DOI: 10.1016/j.ijggc.2012.02.004

[161] Grimm A, Kramer GJ, Gazzani M. How would ideal sorbents improve the technical and economic performance of adsorption-based direct air capture? Energy & Fuels [Internet]. 2024;**38**(19):18781-18799. DOI: 10.1021/acs.energyfuels.4c01156

[162] Nisbet-Jones PBR, Fernandez JM, Fisher RE, France JL, Lowry D, Waltham DA, et al. Is the destruction or removal of atmospheric methane a worthwhile option? Philosophical Transactions of the Royal Society A [Internet]. 2022;**380**(2215):20210108. DOI: 10.1098/rsta.2021.0108

[163] Li Q, Meidan D, Hess P, Añel JA, Cuevas CA, Doney S, et al. Global environmental implications of atmospheric methane removal through chlorine-mediated chemistry-climate interactions. Nature Communications [Internet]. 2023;**14**(1):1-10. DOI: 10.1038/s41467-023-39794-7

[164] Gorham KA, Abernethy S, Jones TR, Hess P, Mahowald NM, Meidan D, et al. Opinion: A research roadmap for exploring atmospheric methane removal via iron salt aerosol. Atmospheric Chemistry and Physics. 2024;**24**(9):5659-5670. DOI: 10.5194/acp-24-5659-2024

[165] Araújo K. The emerging field of energy transitions: Progress, challenges, and opportunities. Energy Research and Social Science. 2014;**1**:112-121. DOI: 10.1016/j.erss.2014.03.002

[166] COP29: Satellites Spot Methane Leaks – but 'Super-Emitters' Don't Fix Them. New Scientist [Internet]. Available from: https://www.newscientist.com/article/2456488-satellites-spot-methane-leaks-but-super-emitters-dont-fix-them/ [Accessed: December 10, 2024]

[167] He TL, Boyd RJ, Varon DJ, Turner AJ. Increased methane emissions from oil and gas following the Soviet Union's collapse. Proceedings of the National Academy of Sciences of the United States of America [Internet]. 2024;**121**(12):e2314600121. DOI: 10.1073/pnas.2314600121

[168] Hafeznia H, Pourfayaz F, Maleki A. An assessment of Iran's natural gas potential for transition toward low-carbon economy. Renewable and Sustainable Energy Reviews. 2017;**79**:71-81. DOI: 10.1016/j.rser.2017.05.042

[169] Madani K. Game theory and water resources. Journal of Hydrology (Amst). 2010;**381**(3-4):225-238. DOI: 10.1016/j.jhydrol.2009.11.045

[170] We Must Go Above, Beyond Compliance, Fully Protect Civilians against 'Harms They Are Suffering on Our Watch', Senior Humanitarian Official Tells Security Council. Meetings Coverage and Press Releases [Internet]. Available from: https://press.un.org/en/2024/sc15702.doc.htm [Accessed: December 10, 2024]

[171] 3 Major Climate Consequences of Trump's Election [Internet]. Available from: https://www.forbes.com/sites/davidcarlin/2024/11/07/donald-trump-and-the-climate-consequences-of-the-2024-us-election/ [Accessed: December 10, 2024]

[172] Massive Development Cooperation Budget Cuts in The Netherlands Damage Our International Position and Own Interests - Partos English [Internet]. Available from: https://www.partos.nl/en/news/massive-development-cooperation-budget-cuts-in-the-netherlands-damage-our-international-position-and-own-interests/ [Accessed: December 10, 2024]

[173] Disaster due to Cuts at University - Background - Utrecht University [Internet]. Available from: https://www.uu.nl/en/background/disaster-due-to-cuts-at-university [Accessed: December 10, 2024]

[174] China Responsible for 95% of New Coal Power Construction in 2023, report says - Carbon Brief [Internet]. Available from: https://www.carbonbrief.org/china-responsible-for-95-of-new-coal-power-construction-in-2023-report-says/ [Accessed: December 10, 2024]

Chapter 2

Perspective Chapter: Environmental Impact and Carbon Management Strategies within Natural Gas Production

Parinaz Soleimani and Ehsan Hakimi

Abstract

Recent advancements in natural gas production prioritize reducing methane emissions, enhancing carbon capture, and innovating flaring reduction techniques. Efforts by organizations like the International Energy Agency focus on converting unused gas into valuable products to mitigate emissions. Methane reduction benefits include lowering the greenhouse gas footprint, improving air quality, and protecting ecosystems. Regulatory compliance, economic benefits, and technological innovation are integral to carbon management strategies. Carbon capture and storage (CCS) is essential for mitigating CO_2 emissions and supporting sustainable development. Innovative flaring reduction techniques also contribute to environmental sustainability, resource conservation, and regulatory compliance. These industry innovations, combined with supportive policies such as regulatory frameworks and incentives, drive substantial reductions in greenhouse gas emissions and promote economic feasibility. This collaborative approach facilitates the transition to a sustainable energy future, supported by public-private partnerships that enhance the development and deployment of innovative solutions, aligned with global emission reduction goals.

Keywords: environmental impact, carbon capture and storage (CCS), methane emission, flaring reduction, industry innovation, policy, greenhouse gas emission

1. Introduction

The global economy relies heavily on oil and gas resources (**Figure 1**). The first commercial natural gas well was drilled in Fredonia, New York, in 1821. By the late nineteenth and early twentieth centuries, advancements in pipeline technology facilitated the widespread distribution and use of natural gas for lighting and heating. The mid-twentieth century saw a significant expansion in natural gas infrastructure and consumption, driven by the development of long-distance pipelines and the

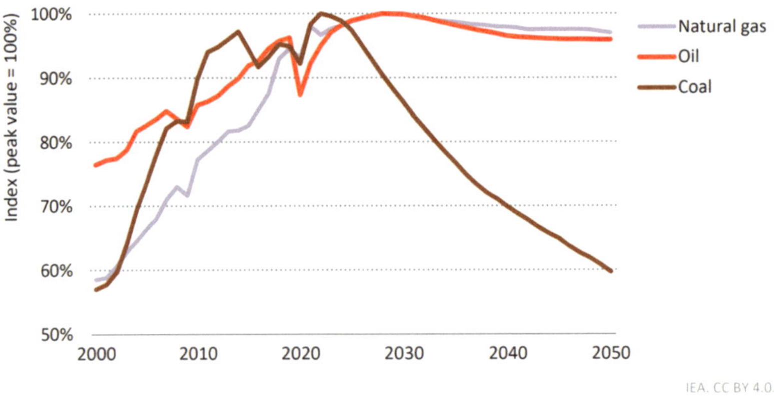

Figure 1.
Fossil fuel consumption by fuel in the STEPS, 2000–2050 [1].

discovery of vast natural gas reserves. The 1970s energy crisis further emphasized the importance of natural gas as an alternative to oil, leading to increased exploration and production activities globally.

Natural gas is a vital component of the global energy mix, accounting for approximately 24% of the world's energy consumption. It is favored for its relatively lower carbon dioxide emissions than coal and oil, making it a key transitional fuel in the shift toward cleaner energy sources. The flexibility and reliability of natural gas also support the integration of renewable energy into the power grid.

However, hydrocarbon exploitation projects can have significant impacts on the environment. Despite its advantages, natural gas production poses significant environmental challenges [2]. Methane, the primary component of natural gas, is a potent greenhouse gas with a global warming potential approximately 25 times greater than carbon dioxide over 100 years. Methane emissions occur throughout the natural gas supply chain, from extraction to transportation and distribution. Additionally, flaring and venting contribute to air pollution and resource wastage.

The environmental impact of natural gas production began to garner attention in the latter half of the twentieth century. Early concerns focused primarily on air and water pollution, as well as land disturbance caused by drilling activities. The Clean Air Act of 1970 in the United States marked one of the first major regulatory efforts to address air pollution from industrial sources, including natural gas production. The 1980s and 1990s saw the introduction of more stringent environmental regulations to reduce emissions from the energy sector. The Clean Water Act, enacted in 1972, and subsequent amendments played a crucial role in regulating the discharge of pollutants into water bodies. During this period, the concept of environmental impact assessments (EIA) became integral to the planning and approval process for new natural gas projects.

The turn of the twenty-first century brought increased awareness of climate change and the role of greenhouse gases (GHGs) in global warming. Methane emissions from natural gas production became a focal point due to methane's high global warming potential. The Kyoto Protocol, adopted in 1997, set binding emission reduction targets for industrialized countries, further driving the need for effective carbon management strategies.

The evolution of carbon management strategies in natural gas production has been driven by the need to mitigate environmental impacts while ensuring economic viability. Early efforts focused on regulatory compliance and reducing operational inefficiencies. Over time, technological advancements and growing awareness of climate change have led to more sophisticated approaches, including methane detection and capture technologies (CCS), and innovative flaring reduction techniques. These strategies reduce emissions and enhance resource efficiency and economic performance. In addition, in recent years, global warming and climate change have caused the world to take special notice of the GHGs and try to find the roots of them and find the best solutions for these problems.

2. Recent advancements in methane emission reduction

Methane emissions make the second largest contribution to global greenhouse gas (GHG) radiative forcing, accounting for approximately 18%—not including indirect effects—compared to 75% from CO_2 [3]. Methane emissions originate from natural sources (e.g., Wetlands, the oceans, termites), agriculture (e.g., cattle, rice fields), biogas, waste, and the incomplete combustion of fossil fuels [4]. Approximately 10–15% of annual, global methane emissions arise from oil and gas supply chains and use [4].

Current efforts to mitigate methane emissions have led to significant advancements, particularly in the gas industry. Key strategies in these efforts include:

- A leak detection and repair (LDAR) program,
- Improving venting and flaring practices,
- Upgrading equipment and infrastructure.

2.1 A leak detection and repair (LDAR) program

Leak detection and repair (LDAR) programs are designed to identify, quantify, and mitigate leaks of volatile organic compounds (VOCs), including methane, a potent greenhouse gas with a global warming potential many times higher than carbon dioxide over 20 years.

Traditionally, LDAR programs relied on manual inspections and portable monitoring devices to detect leaks. However, technological advancements have significantly improved these efforts, making leak detection more efficient and accurate. Today, these programs use advanced technologies such as infrared cameras, drones, and satellite monitoring to detect methane leaks in infrastructure. Combining regular maintenance and rapid repair protocols with these advanced technologies has greatly reduced fugitive emissions and enhanced management practices.

Some of the notable technologies used in modern LDAR programs include:

- Infrared cameras: Infrared cameras, also known as optical gas imaging (OGI) cameras, are one of the most widely used technologies in modern LDAR programs. These devices can visualize methane leaks in real time by detecting the infrared radiation emitted by methane molecules. They are widely used for routine inspections and leak detection at production sites. These cameras are portable and can be used to scan large areas, making them particularly useful for on-site inspections in refineries, pipelines, and other industrial facilities.

- Laser-based sensors: Tunable diode laser absorption spectroscopy (TDLAS) is a technique used to measure methane concentrations accurately. These sensors can be deployed on drones or installed in fixed locations to monitor emissions continuously.

- Satellite-based monitoring: Satellites equipped with hyperspectral imaging sensors can detect methane emissions from space. Programs like the European Space Agency's Copernicus Sentinel-5P and Greenhouse Gases (GHGS) provide valuable data on large-scale methane emissions. This technology is particularly useful for tracking emissions in remote or politically sensitive areas where ground-based inspections are not feasible.

- Mobile leak detection units: These units are equipped with high-precision sensors and can be mounted on vehicles to conduct comprehensive surveys of methane leaks across production sites and pipelines.

2.2 Improving venting and flaring practices

In the oil and gas industry, venting and flaring are common practices used to manage excess gases that are either not economically viable to capture or pose a safety risk if allowed to accumulate. Venting involves the direct release of gas into the atmosphere while flaring involves burning the gas, converting it into carbon dioxide (CO_2) and water vapor. According to a recent assessment, flaring and venting waste 8% of global natural gas production annually, contribute 6% of global greenhouse gas emissions, and disperse a range of other harmful pollutants. Upstream natural gas flaring includes operational and safety reasons and flaring for economic reasons. Flaring for operational and safety reasons is generally short term. The flaring of relatively large volumes of gas associated with oil production, either temporarily or long term, is the area that has generated the most concern among stakeholders [5].

Recent advancements have focused on converting flared gas into valuable products through technologies such as gas-to-liquids (GTL) and microchannel reactors [6, 7]. Here is a brief explanation of these technologies:

2.2.1 Gas-to-liquids (GTL)

This process converts captured methane into liquid fuels or chemicals through catalytic reactions, offering a valuable product while simultaneously reducing emissions. The versatility of the output—ranging from diesel and jet fuel to lubricants and chemicals—enhances the effectiveness of gas-to-liquid (GTL) technology by creating multiple revenue streams.

Setting up GTL plants requires significant capital investment in infrastructure and technology. Operational costs also include the need for catalysts and energy to drive the conversion process. Setting up GTL is often economic when natural gas prices are low and there is a high demand for liquid fuels. GTL technology allows for monetizing stranded gas reserves that would otherwise be flared, converting them into transportable and commercially valuable products. It also has potential applications in producing chemicals and other industrial products.

This transformation is achieved through a chemical process known as Fischer-Tropsch synthesis, which converts synthesis gas (a mixture of carbon monoxide and hydrogen) into liquid hydrocarbons. Under optimized conditions, this process can achieve conversion efficiencies ranging from 60% to 70%. Companies like Shell and Sasol have developed large-scale GTL plants, demonstrating the commercial viability and environmental benefits of this technology.

The process involves a series of chemical reactions that convert synthesis gas into liquid hydrocarbon derivatives. These reactions take place in the presence of metal catalysts, typically at temperatures between 150°C and 300°C (300–570°F) and pressures ranging from 15 to 500 psi, using a variety of catalysts (**Table 1**).

Product distributions are influenced by temperature, feed gas composition (H_2/CO), pressure, catalyst type, and catalyst composition. Fischer-Tropsch products are produced in four main steps: (i) synthesis gas generation, (ii) gas purification, (iii) Fischer-Tropsch synthesis, and (iv) product upgrading. Depending on the desired types and quantities of Fischer-Tropsch products, either low-temperature (200–240°C; 390–465°F) or high-temperature (300–350°C; 570–660°F) synthesis is used, utilizing either iron (Fe) or cobalt (Co) catalysts.

The required mixture of carbon monoxide and hydrogen (synthesis gas) is generated through a reaction involving gas, water steam, and oxygen at temperatures above 900°C. The synthesis gas generation, coupled with Fischer-Tropsch hydrocarbon synthesis, forms a two-stage sequence of reactions that enables the production of liquid fuels like diesel and gasoline.

The Fischer-Tropsch synthesis process essentially builds carbon chains by attaching methylene groups to a carbon chain. These reactions are highly exothermic, releasing a significant amount of heat. To maintain stable reaction conditions and prevent the formation of lower-boiling hydrocarbon derivatives, adequate cooling is essential. The total heat released by the reaction is approximately 25% of the heat of combustion of the synthesis gas, which sets a theoretical limit on the maximum efficiency of the Fischer-Tropsch process.

Catalysts play a crucial role in the Fischer-Tropsch process, with iron and cobalt being the most used. The choice of catalyst depends on the desired product spectrum and operating costs. Iron-based catalysts, for example, are used in all commercial Sasol Fischer-Tropsch reactors. Cobalt-based catalysts, known for their high conversion rates at low temperatures, are not suitable for high-temperature applications due to excessive methane formation. Despite its high cost, cobalt is favored for maximum diesel production in a once-through process, and Sasol has developed cobalt catalysts that perform well in slurry phase processes.

Other catalysts, such as nickel-based and ruthenium-based options, also exhibit enough activity for commercial use. However, their practical applications are limited: ruthenium is scarce, and nickel, while highly active, produces too much methane and performs poorly at high pressures due to the formation of volatile carbonyls. This makes cobalt and iron the most practical choices for catalysts in GTL processes.

Catalyst	Reaction type	Products
Ni	Methanation	Methane, CH_4
Cu/Cr/Zn oxide	Methanol synthesis	Methanol, CH_3OH
Fe, Co	Fischer-Tropsch synthesis	Hydrocarbons, C_nH_{2n+2}

Table 1.
Summary of the catalysts for the conversion of synthesis gas to products.

Gas-to-liquids (GTL) technology has seen significant advancements in recent years, driven by the need to reduce flaring, such as improved catalysts and processes, which focus on developing more effective catalysts that increase conversion rates and selectivity toward desired liquid hydrocarbons. These new catalysts are designed to operate under milder conditions, reducing energy consumption and operational costs and higher-purity diesel, naphtha, and lubricants. These high-quality products meet stringent environmental regulations and market requirements, expanding the commercial viability of GTL products [8].

The development of modular GTL plants has revolutionized the industry. These small-scale, transportable plants can be deployed in remote locations, allowing for the economic conversion of associated gas that would otherwise be flared. Modular plants offer flexibility and scalability, making them ideal for onshore and offshore applications [9].

Also, integrating GTL technology with renewable energy sources such as solar and wind power has emerged as a promising approach to further reduce the carbon footprint of the process. This integration helps provide a steady and sustainable energy supply for GTL operations, enhancing the overall environmental benefits of the technology.

In addition, the adoption of digital technologies and automation in GTL plants has significantly improved operational efficiency and reliability. Advanced monitoring and control systems enable real-time optimization of the GTL process, reducing downtime and maintenance costs. Digital twins and predictive analytics further enhance the performance and longevity of GTL facilities.

Finally, recent projects, such as the development of large-scale GTL plants by Shell and Sasol, have demonstrated the commercial viability of GTL technology. The ability to produce high-value products from natural gas enhances the economic attractiveness of GTL technology.

These innovations contribute to a more sustainable and profitable natural gas sector, aligning with global efforts to reduce greenhouse gas emissions and promote cleaner energy alternatives. It also greatly helps their ability to remain in the energy market and easily and quickly adapt to environmental and carbon management laws, policies, and standards.

2.2.2 Microchannel reactors

Microchannel reactors are compact devices that enhance the efficiency of chemical reactions by reducing the diffusion distances of reactants. These reactors can convert natural gas into various chemicals and fuels at lower temperatures and pressures compared to traditional reactors. These reactors contain numerous parallel microchannels, each with a diameter typically in the range of millimeters, which significantly increases the surface area-to-volume ratio. This enhancement leads to improved heat and mass transfer rates, enabling faster and more efficient chemical reactions.

The use of microchannel reactors in flare gas applications has shown promise in capturing and converting small volumes of gas at production sites, reducing emissions, and creating value from what would otherwise be waste.

Recent advancements have focused on optimizing the design and scalability of these reactors. For instance, the numbering-up approach allows for the scaling of reaction output without expanding the reactor dimensions. By running multiple microchannel reactors in parallel within a single module, the overall production capacity can be increased while maintaining efficiency and control over reaction conditions.

Furthermore, microchannel reactors are being integrated into small-scale and modular GTL plants. These plants are designed to be highly flexible and can be deployed in remote locations, making them ideal for capturing and converting flare gas at production sites. This modularity not only reduces transportation and installation costs but also allows for easy scalability based on the availability of gas.

Despite their advantages, microchannel reactors face several challenges. One major issue is the high initial cost associated with the development and deployment of these reactors. The precise fabrication of microchannels and the integration of advanced catalysts require significant investment.

Another challenge is the management of heat within microchannels. Given the high surface area-to-volume ratio, maintaining uniform temperature distribution is critical to prevent hotspots that could degrade catalysts or cause undesired side reactions. Advanced cooling systems and precise temperature control mechanisms are essential to address this issue.

Moreover, microchannel reactors' long-term durability and maintenance in harsh operational environments pose additional challenges. Continuous research and development are needed to enhance the robustness and lifespan of these reactors under varying conditions [10].

However, microchannel reactors offer several environmental benefits, particularly in reducing emissions from flaring. This helps mitigate climate change and transforms waste gas into valuable products, enhancing resource efficiency.

The high efficiency of microchannel reactors also translates into lower energy consumption compared to traditional reactors. Additionally, the compact design of these reactors minimizes land use and associated environmental disturbances.

2.2.3 Other innovations and technologies

Gas capture and utilization technologies have been developed to reduce flaring and its environmental impact. Flare Gas Recovery Systems (FGRS) capture gas that would otherwise be flared and repurpose it for power generation or re-injection into the production process. The U.S. Department of Energy (DOE) actively supports these innovations through funding and research initiatives aimed at reducing methane emissions and converting flared gas into commercially viable products [5]. Companies like Honeywell and Siemens offer commercially available FGRS solutions that have been successfully implemented in various oil and gas operations.

Improving operational efficiencies is another critical approach to reducing flaring. This involves better planning and coordination of drilling and production activities to ensure adequate gas takeaway capacity. Effective planning and coordination of drilling and production activities are essential for minimizing flaring. This involves scheduling well completions and production ramp-ups to coincide with the availability of gas takeaway infrastructure, such as pipelines or processing facilities. By aligning production schedules with transportation capacity, operators can reduce instances where gas must be flared due to a lack of available infrastructure. For example, ExxonMobil's initiatives in the Permian Basin have demonstrated that zero routine flaring is achievable through enhanced operational protocols, such as installing new capture equipment and ensuring wells are completed only when there is sufficient gas transportation infrastructure.

Utilization of gas compression and re-injection techniques is another method. It involves capturing associated gas at the wellhead, compressing it, and re-injecting it into the reservoir or a nearby formation for storage or enhanced oil recovery (EOR).

This method not only reduces flaring but also enhances oil production by maintaining reservoir pressure. Advanced compression technologies and high-capacity compressors are critical for efficiently handling the volumes of gas produced during oil extraction.

Installing new equipment designed to capture and repurpose emissions is a proactive approach to reducing flaring. This includes vapor recovery units (VRUs), which capture vapors from storage tanks and production facilities and redirect them for use in the production process or sale. Additionally, mobile gas processing units can be deployed to remote or temporary production sites to capture and process flared gas on-site, providing flexibility and reducing environmental impact.

2.3 Upgrading equipment and infrastructure

Retrofitting aging infrastructure with more efficient and low-emission equipment is a critical strategy for reducing methane emissions in the oil and gas industry. However, the challenges and limitations associated with retrofitting, including high costs, technical complexity, and operational disruptions, must be carefully managed to ensure the success of these efforts. As regulatory pressures increase and the economic benefits of emission reductions become more apparent, the adoption of retrofitting technologies is likely to grow, contributing to a more sustainable and environmentally responsible natural gas industry. Technologies such as advanced pneumatic controllers, compressors, new pipeline materials, and redesigning pipelines offer significant potential for improving the environmental performance of gas operations. Here are more details of these methods:

- Pneumatic controllers, traditional pneumatic controllers, often use natural gas to power their operations, which can result in significant methane emissions if the gas is vented into the atmosphere. Retrofitting these controllers with low-emission or zero-emission alternatives, such as electric or instrument air-powered controllers, can drastically reduce methane emissions. These modern controllers operate with greater precision and reliability, further contributing to overall efficiency in gas management systems.

- Compressors can drastically cut methane emissions. Older compressor systems often suffer from leaks and inefficiencies that lead to methane emissions. Modifying these systems with modern compressors through improved seals, advanced materials, and more efficient operations controls designed to minimize fugitive emissions can reduce emissions.

- New materials, reconstructing existing pipelines with advanced materials such as composite materials or corrosion-resistant alloys, can significantly enhance their durability and leak resistance. These new materials are designed to withstand the harsh conditions often encountered in natural gas operations, such as high pressures, extreme temperatures, and corrosive environments.

- Designs for pipelines. For example, pipelines can be designed with thicker walls, more robust joints, and enhanced protective coatings to prevent corrosion and mechanical damage. Additionally, modern pipelines often include built-in sensors and monitoring systems that can detect pressure drops or changes in gas composition, which may indicate a leak. These design improvements make pipelines more resilient to both internal and external stresses, thereby reducing the likelihood of methane emissions.

- Storage facilities, redesigning and modifying these facilities with modern, low-emission equipment such as high-performance seals and valves, can greatly reduce the risk of leaks. Additionally, the use of advanced monitoring systems that continuously track the integrity of storage facilities can help operators detect and address potential issues before they result in significant emissions.

3. Methane capture and utilization technologies

Methane capture and utilization technologies involve capturing methane emissions at various supply chain stages from wells to users and converting them into useful products such as liquefied natural gas (LNG) or chemicals. These technologies reduce emissions and provide economic benefits by creating additional revenue streams. These aspects of benefits could encourage industries to change their strategies and try to achieve these technologies.

There are several technologies and techniques for capturing and utilizing methane. Key technologies and methods include vapor recovery units (VRUs), adsorption technologies, methane-to-liquids (MTL) technology (explained in Section 2.2.1), Biogas upgrading, Electrochemical conversion, and Biological methane conversion. Below is a concise explanation of each:

3.1 Vapor recovery units (VRUs)

They are crucial systems employed in the natural gas industry to manage and recover vapors that are released during the storage and processing of natural gas. The primary purpose of VRUs is to minimize the release of volatile organic compounds (VOCs) and other pollutants into the atmosphere, thereby reducing environmental impact and improving air quality. These units are designed to capture methane emissions from storage tanks and other equipment, which not only helps in reducing environmental pollution but also provides economic benefits by recovering valuable hydrocarbons. The process involves drawing hydrocarbon vapors out of storage tanks, separating any condensed liquids, and then compressing and directing the vapors to sales lines or for on-site use. Key components of a VRU may include:

1. Vapor capture system: A system designed to collect vapors from the source.

2. Compressors: To increase the pressure of the captured vapors for further processing or transport.

3. Condensers: For cooling the vapors, which can lead to the condensation of recoverable liquids.

4. Storage tanks: For storing recovered liquids before they are returned to the system or further processed.

5. Control systems: To monitor and manage the operation of the VRU to ensure efficiency and safety.

The environmental benefits of using vapor recovery units (VRUs) are significant. Primarily, VRUs contribute to the reduction of methane emissions, a potent greenhouse gas, by capturing it from storage tanks and other sources in the natural gas industry. This process helps mitigate the impact of climate change. Additionally, VRUs control emissions of volatile organic compounds (VOCs) and hazardous air pollutants (HAPs), which are known to contribute to air pollution and can have adverse health effects. By reducing the release of these toxic contaminants, VRUs improve air quality and align with environmental compliance requirements. Furthermore, the recaptured vapors, rich in natural gas liquids (NGLs) and other hydrocarbons, can be repurposed as a fuel source, enhancing fuel efficiency and reducing the need for new fossil fuel extraction.

Also, vapor recovery units (VRUs) have a notable economic impact. Recovered gas can be redirected into sales lines or used on-site, which can lead to significant cost savings and additional revenue streams. The efficiency of VRUs in capturing about 95% of hydrocarbon vapors translates into a direct economic benefit, as these vapors are rich in natural gas liquids (NGLs) and have a higher heat content than pipeline-quality natural gas. Consequently, the use of VRUs can result in substantial returns, especially in markets where there is demand for the recovered, energy-rich vapors. Moreover, the relatively low cost of VRU technology compared to the value of the recovered gas makes it a financially sound investment for companies looking to enhance their operational efficiency and environmental stewardship. This technology is relatively easy to integrate into existing facilities in the upstream, midstream, and downstream sectors of the oil and gas industry, and it can be designed to handle various types of hydrocarbons. The installation and maintenance processes are well-understood and supported by numerous service providers.

The keys to cost-effective VRU projects are a steady source and adequate quantity of sources of vapors, along with an economic outlet for the collected products. The potential volume of vapors will depend on the makeup of the collection system and the quantity and types of sources it is connected to. Pipeline connection costs for routing vapors off-site should be considered when selecting sites for VRU installation.

3.2 Adsorption technologies

Advanced materials, such as activated carbon, zeolites, and metal-organic frameworks (MOFs), are being developed to adsorb methane at low concentrations, enabling its capture from dilute emission streams. These materials have high surface areas and can selectively adsorb methane at certain pressures and temperatures.

The process begins with the introduction of a gas mixture (e.g., natural gas or methane-rich biogas) into a vessel containing the adsorbent. Under appropriate conditions, methane molecules preferentially adhere to the adsorbent surface. Once the adsorbent reaches its capacity, the methane can be released (desorbed) by changing conditions (e.g., reducing pressure, increasing temperature) so it can be collected and utilized.

One of the most widely used adsorption technologies in gas separation is pressure swing adsorption (PSA), which takes advantage of changing pressure to alternate between adsorption and desorption cycles. It uses pressure changes to adsorb and desorb methane, making it a highly efficient method for continuous operation. This technology can efficiently separate methane from other gases like CO_2, nitrogen, or hydrogen sulfide (like natural gas processing), enabling its capture and further utilization. It is highly scalable, making it suitable for small and large operations. It can be implemented in small biogas plants to large natural gas processing facilities. The modular nature of adsorption systems allows for easy scaling up or down depending on the volume of gas to be processed. Additionally, adsorption systems can be designed to operate safely, with a lower risk of fire or explosion compared to other gas handling methods.

3.3 Biogas upgrading

Methane captured from anaerobic digestion processes at landfills or wastewater treatment plants can be upgraded to pipeline-quality natural gas using membrane separation or pressure swing adsorption (PSA) technologies.

Also, studying in this field led to the introduction of emerging technologies that offer promising solutions for methane emission reduction. Here are two of them:

3.3.1 Electrochemical conversion

Cutting-edge research focuses on electrochemical methods for converting methane into valuable chemicals at ambient temperatures. Metal oxide electrocatalysts, for example, have shown potential in efficiently transforming methane into methanol and other oxygenates. These processes can operate under mild conditions, offering a sustainable alternative to traditional high-temperature catalytic methods.

3.3.2 Biological methane conversion

Innovations in microbial and enzymatic pathways are being explored to biologically convert methane into methanol and other chemicals. These biological systems can potentially operate at lower energy costs and with higher specificity in contrast with conventional chemical processes.

Besides all the advantages and utilization of these technologies, there are still some significant challenges. One of the most important challenges is the cost of implementation of them. Most are still not commercial and not cost-effective for middle and small companies. As mentioned before, some of these technologies like biogas upgrading and biological methane conversion are very new and need more studies to be more adaptable and practical for companies.

4. Carbon capture, utilization, and storage (CCUS)

Carbon capture, utilization, and storage (CCUS) encompasses methods and technologies to remove CO_2 from the flue gas and the atmosphere, followed by recycling the CO_2 for utilization and determining safe and permanent storage options.

Ideally, converting CO_2 into useful chemicals of commercial importance, or utilizing CO_2 for oil extraction or modification of alkaline industrial wastes, would add economic value to this greenhouse gas. However, the demand for CO_2 is limited in comparison to the vast amount of CO_2 that needs to be removed from the atmosphere, to reduce the harmful environmental impacts of climate change. Therefore, various options for CO_2 storage have been proposed. These options include injecting CO_2 in geologic formations and oceans, and growing trees to enable biological fixation of CO_2 via photosynthesis. Any stable system for storing carbon must be effective and cost-competitive, stable as long-term storage, and environmentally safe [11].

The oil and gas industry is already one of the global leaders in developing and deploying carbon capture and utilization (CCUS) technologies. Of the 35 Mt CO_2 captured today from industrial activities in large-scale CCUS facilities, nearly 80% is captured from oil and gas operations (**Figure 2**) [12].

4.1 Carbon capture and storage (CCS) technologies and mechanisms

There are several technologies used in CCS, including pre-combustion capture, post-combustion capture, and oxy-fuel combustion. Each of these technologies has its mechanisms and applications, depending on the specific requirements of the emission source.

- Pre-combustion capture: Pre-combustion capture involves the gasification of fossil fuels to produce a synthesis gas (syngas), which is then subjected to a water-gas shift reaction to increase the proportion of hydrogen (H_2) and CO_2, facilitating the separation of CO_2 before combustion occurs. This process is generally conducted in three main steps: gasification, water-gas shift reaction, and CO_2 separation. In the gasification step, fossil fuel, such as coal, natural gas, or biomass, is reacted with oxygen (O_2) or steam (H_2O) at high temperatures to produce a synthesis gas (syngas). The syngas is primarily composed of hydrogen (H_2), carbon monoxide (CO), and some CO_2. Then in the second step, the syngas is subjected to a water-gas shift reaction, where carbon monoxide (CO) reacts with steam (H_2O) to produce more hydrogen (H_2) and carbon dioxide (CO_2). This reaction increases the CO_2 content in the gas mixture, making it easier to separate CO_2 from hydrogen. Finally, in the last step, the CO_2 is separated from the hydrogen using various techniques, such as chemical absorption, physical absorption, or membrane separation. The resulting pure CO_2 can then

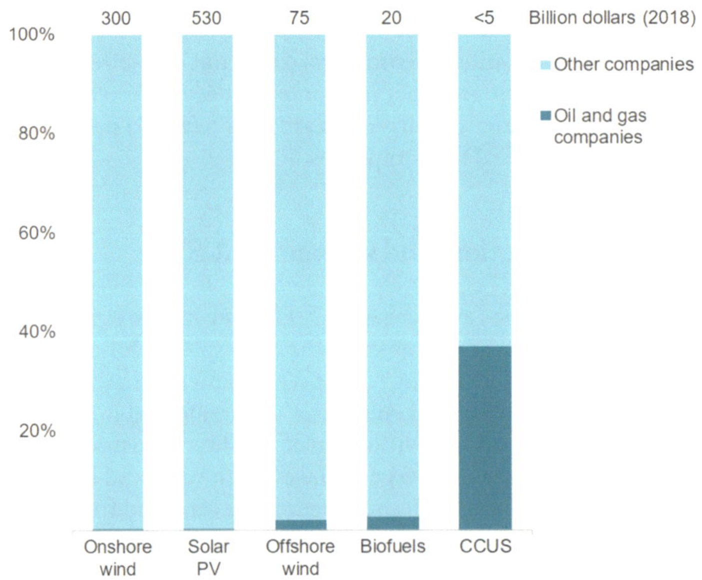

Figure 2.
Global capital investment in selected low-carbon technologies (2015–2018). Note: CCUS only includes large-scale facilities [12].

be captured, compressed, and transported for storage. This method is highly efficient because it captures CO_2 before combustion, resulting in a higher concentration of CO_2 in the gas stream, which makes the separation process more straightforward. This method produces hydrogen as a by-product, which can be used as a clean fuel for power generation or in other industrial processes, contributing to the decarbonization of various sectors. Besides these benefits, it has some limitations. It needs specific equipment and conditions, making it complex and expensive to implement on a large scale. Also, it often requires new infrastructure, as it is not easily modified to existing power plants.

- Post-combustion capture: This method captures CO_2 from flue gases after fossil fuels have been burned, typically using solvent-based chemical absorption. It is suitable for retrofitting existing power plants. Now, we focus more on this method. First, after combustion, the flue gases, which contain CO_2, nitrogen (N_2), water vapor (H_2O), and other minor components, are cooled and treated to remove pollutants like sulfur oxides (SO_x) and nitrogen oxides (NO_x) before entering the CO_2 capture unit. Then, the cooled flue gas is passed through an absorber column, where it meets a solvent, typically an amine-based solution, such as mono ethanol amine (MEA). The CO_2 in the flue gas chemically reacts with the amine to form a carbamate compound. Finally, the captured CO_2 is compressed and prepared for transportation to storage sites, where it can be injected into geological formations. It can be added to existing power plants and industrial facilities without requiring major changes to the existing infrastructure, making it a practical option for reducing emissions from current sources, and chemical absorption, particularly using amine-based solvents, is a well-established and commercially available technology, with significant research and operational experience backing its use. The solvents used in chemical absorption can degrade over time, especially in the presence of impurities like SO_x and NO_x, leading to higher operational costs and the need for regular solvent replacement. Scaling up this method shows challenges in managing the significant volumes of flue gas that must be processed.

- Oxy-fuel combustion: This technique burns fossil fuels in oxygen instead of air, resulting in a flue gas that is primarily CO_2 and water vapor, which makes the capture process easier. In the first step, oxygen is produced through air separation units (ASUs), which typically use cryogenic distillation to separate O_2 from nitrogen (N_2) in the air. Second, the fossil fuel is burned in an environment of pure oxygen, rather than air, to produce a flue gas composed mainly of CO_2 and water vapor (H_2O). The absence of nitrogen in the combustion process prevents the formation of nitrogen oxides (NO_x), reducing the need for flue gas treatment. The flue gas, mainly consisting of CO_2 and H_2O, is cooled to condense the water vapor, leaving behind a highly concentrated stream of CO_2. The CO_2 is then compressed, purified if necessary, and transported for storage. This method produces a flue gas that is rich in CO_2, which significantly simplifies the capture process and reduces the cost associated with CO_2 separation. Additionally, eliminating nitrogen from the combustion process minimizes the production of NO_x, reducing the need for post-combustion treatment of flue gases. On the other hand, the need for specialized equipment, such as ASUs and flue gas processing units, increases the capital and operational costs associated with oxy-fuel combustion and, while retrofitting is possible, the high costs and technical challenges associated with integrating oxy-fuel combustion into existing power plants may limit its widespread adoption [13].

4.2 Implementation cost of CCS

The implementation cost of CCS technologies is influenced by several factors, including the cost of implementation, the type of technology used, the scale of the project, the location of the storage site, policy measures and regulatory frameworks, research and development, and public-private partnerships.

- Capital costs: These include expenses related to the construction of capture facilities, transportation infrastructure, and storage sites. For instance, building a large-scale post-combustion capture facility can cost between $500 million and $1 billion. Oxy-fuel combustion and pre-combustion capture technologies may have different cost structures due to their specific requirements.

- Operational costs: These include the costs associated with running the capture and storage facilities, such as energy consumption, maintenance, and monitoring. Operating costs for CCS are estimated to range from $30 to $70 per ton of CO_2 captured, depending on the efficiency of the technology and the characteristics of the emission source.

- Storage costs: The cost of storing CO_2 depends on the geology of the storage site and the distance from the capture location. Storing CO_2 in depleted oil and gas fields tends to be more expensive than in saline aquifers. Storage costs can range from $10 to $20 per ton of CO_2.

- Economic incentives and policy impact: The cost-effectiveness of CCS is significantly influenced by economic incentives and policy frameworks. Governments may provide financial incentives, such as tax credits, grants, or subsidies, to offset some of the capital and operational costs. For example, the 45Q tax [2] credit in the United States offers up to $50 per ton of CO_2 stored, which can make CCS projects more economically viable. The European Union Emissions Trading System (EU ETS) has been instrumental in reducing greenhouse gas emissions across Europe. Putting a carbon price on carbon incentivizes companies to adopt cleaner technologies and practices. The system has supported the deployment of CCS and methane reduction technologies in various industries, including natural gas production [14]. Some jurisdictions mandate the use of CCS for certain industries or power plants, particularly those with high CO_2 emissions. These regulations can drive the adoption of CCS technologies by creating a legal requirement for emission reductions.

If the incentives are in place to encourage investment in different components of the CCUS value chain, there are several ways to think about the role of the oil and gas industry concerning CCUS:

- As a source of some concentrated streams of CO_2 that are relatively easy and cost-effective to capture (for example, in gas processing or parts of the refining sector).

- As a user of CO_2, primarily for injection into reservoirs as a mechanism for Enhanced oil recovery (EOR). Depending on the source of the CO_2 and the volumes being injected, this could reduce the emissions intensity of the produced oil considerably (and even, theoretically, lead to carbon-negative oil).

- As an industry that undertakes well-funded, high-level research, and that has large-scale engineering, pipeline and subsurface, and project management capabilities to scale up CCUS. This could have positive spillover implications for many aspects of energy transitions, including for the large-scale production of low-carbon hydrogen and the decarbonization of heavy industry [12].

- Carbon pricing: Implementing carbon pricing mechanisms, such as carbon taxes or cap-and-trade systems, can incentivize companies to adopt CCS by increasing the cost of emitting CO_2. A higher carbon price makes CCS more attractive by improving its economic competitiveness relative to paying for emissions.

- Research and development funding: Government funding for R&D can accelerate the development of advanced CCS technologies, reducing costs and improving efficiency. Public investment in pilot projects and demonstration plants helps to de-risk the technology and pave the way for commercial-scale deployment.

- Public-private partnerships: Collaborative efforts between governments, private companies, and research institutions can facilitate the deployment of CCS. These partnerships often provide financial support, share technical expertise, and promote the commercialization of new technologies.

Partnerships are an integral part of the oil and gas industry operation. The industry tends to share risks in areas with higher costs, which is why there are typically many partners in large offshore oil fields or LNG terminals. It has also created numerous bodies to deal with industry-wide issues of environmental performance, such as the Oil and Gas Climate Initiative (OGCI).

Existing partnerships and associations can promote some strategic responses to these risks. For example, methane abatement is a focus for the OGCI and the signatories to the multi-stakeholder Methane Guiding Principles group.

- International collaboration: Global cooperation on CCS, through initiatives such as the Carbon Sequestration Leadership Forum (CSLF) and Mission Innovation, helps to standardize best practices, share knowledge, and leverage resources for large-scale deployment.

4.3 Case studies of CCS in natural gas production

Several case studies illustrate the successful implementation of CCS technologies in natural gas production. These projects demonstrate the feasibility, challenges, and benefits of CCS in reducing CO_2 emissions.

- Sleipner CO_2 Storage Project, Norway:

  The Sleipner project, operated by Equinor (formerly Statoil), is one of the world's first commercial CCS projects. Since 1996, CO_2 has been captured from natural gas production at the Sleipner field and injected into the Utsira sandstone formation beneath the North Sea. This project has successfully stored over 20 million tons of CO_2, demonstrating the long-term stability and effectiveness of geological storage.

- Petra Nova Project, USA:The Petra Nova project, a joint venture between NRG Energy and JX Nippon Oil & Gas Exploration, is one of the largest post-combustion carbon capture projects in the world. Located at the W.A. Parish Generating Station in Texas, the project captures CO_2 from a coal-fired power plant and uses it for enhanced oil recovery (EOR) in the West Ranch oil field. Since its inception in 2017, Petra Nova has captured and utilized over 1.6 million tons of CO_2 annually.

- Quest CCS Project, Canada:Operated by Shell, the Quest project captures CO_2 from the Scotford Upgrader, which processes oil sands bitumen into synthetic crude oil. The captured CO_2 is transported via pipeline to a storage site in Alberta, where it is injected into a deep saline aquifer. Since 2015, Quest has successfully captured and stored over 5 million tons of CO_2, providing valuable insights into the scalability and safety of CCS technology.

- Gorgon CO_2 Injection Project, Australia:The Gorgon project, led by Chevron, is one of the largest CCS projects associated with natural gas production. CO2 is captured from the Gorgon LNG plant on Barrow Island and injected into a deep reservoir beneath the island. The project aims to reduce CO_2 emissions by 40% over its lifespan, potentially storing up to 100 million tons of CO_2.

These case studies highlight the technical and economic viability of CCS in various contexts. They also underscore the importance of regulatory support, public-private partnerships, and technological innovation in the successful deployment of CCS projects.

5. Benefits and challenges

5.1 Benefits

Reducing methane and carbon emissions has significant economic benefits. For instance, Companies can monetize resources of flaring that would otherwise be wasted, by capturing and repurposing flared gas. This improves the economic viability of natural gas operations and supports broader energy security and supply chain resilience. Implementing flaring reduction technologies can result in cost savings by reducing the need for flaring permits and compliance with environmental regulations. Additionally, the sale of captured gas and its by-products can offset operational costs.

Generally, if we can convert methane and CO_2 into valuable products that can be sold in the market with benefits, it will be interesting for gas industries to notice the quantity of CO_2 and methane emissions.

From an environmental perspective, methane, the primary component of natural gas, has a global warming potential approximately 25 times greater than carbon dioxide over 100 years. The natural gas industry can substantially lower its GHG footprint by minimizing methane emissions. Improved air quality and reduced health risks for nearby communities are additional environmental advantages of these efforts.

Reducing methane and carbon emissions aligns with sustainable development goals by promoting responsible resource utilization and reducing the environmental footprint of natural gas production. It also supports the transition to a low-carbon economy by providing cleaner energy alternatives.

This aspect of methane and carbon emissions is essential and noticeable for gas industries. In recent years, national and international laws and regulations have focused on it, and if they have more emissions than the standards, they do not let these plants work and be active.

5.2 Challenges

Government policies and regulatory frameworks play a crucial role in driving the adoption of CO_2 emission reduction and carbon management strategies. International agreements such as the Paris Agreement set global targets for emission reductions, encouraging countries to implement stringent regulations. National regulations vary, with some countries imposing carbon pricing mechanisms, emission caps, and incentives for clean technology adoption. The DOE's Methane Emissions Reduction Action Plan, for instance, sets ambitious targets for cutting methane emissions by 30% from 2020 levels by 2030 [5, 15]. This plan includes support for technological innovation and the implementation of best practices across the industry. For example, the United States offers the 45Q tax credit [16] for carbon capture and storage projects, providing financial incentives for companies to implement CCS technologies. But these efforts are not sufficient. In most countries with a big portion of CO_2 and methane emissions, there are a few laws about these subjects. In addition, some laws are not clear enough for industries, and in some cases, these laws are contrary to international laws and standards.

Public-private partnerships (PPPs) are essential for advancing carbon management strategies. PPPs facilitate the development and deployment of innovative technologies by pooling resources, sharing risks, and leveraging expertise from different sectors. These collaborations accelerate the commercialization of new technologies and enhance their scalability, contributing to more effective carbon management in natural gas production. Governments also have a prominent role in this subject because they can support and gather research departments and industries with good coordination. One notable example is the Carbon Capture and Storage Association (CCSA) in the UK, which brings together industry, government, and academic stakeholders to promote CCS deployment. Another example is the Global Methane Initiative, a partnership involving governments and private sector entities to reduce methane emissions worldwide.

Despite all the promotions in mitigating CO_2 and methane emissions, there is still a need for more investment in this field, especially in research for finding a more adaptable and economical solution. In addition, governments can enhance the bankability of these projects, giving developers and investors more confidence. This requires a range of measures, including increased government funding, a well-regulated carbon credit compliance market, and progressive regulations that penalize industries lagging in emission reductions.

As we mentioned before, many of the present solutions are very expensive and not affordable for many countries, especially for underdeveloped countries and small industries. This can be done by financially supporting research, academic centers, and universities that work on these subjects. It can also be done by some international institutes by offering special scholarships in these cases.

6. Conclusion

Methane emission reduction and carbon management within the natural gas industry have seen remarkable advancements over recent years, reflecting a concerted effort to mitigate the environmental impact of fossil fuel utilization.

The continuous evolution of the natural gas industry underscores the paramount importance of environmental impact mitigation and advanced carbon management strategies.

Methane contributes significantly to global warming as a potent greenhouse gas, making its reduction imperative for achieving climate goals.

Current strategies and emerging technologies drive the industry toward more sustainable practices, addressing both methane and CO_2 emissions through a multifaceted approach. The integration of cutting-edge technologies and innovative approaches in methane emission reduction, carbon capture and storage (CCS), and flaring reduction highlights the industry's commitment to sustainability and regulatory compliance.

The collective efforts of industry stakeholders, regulatory bodies, and technological innovators have paved the way for substantial progress in reducing greenhouse gas (GHG) emissions and promoting resource efficiency.

However, this industry has a long way to go to maintain and improve its position in the future as one of the energy producers in the world that has the largest share of the energy market. Despite all the efforts studies and investigations done to reduce the environmental effects of gas production and use different methods to reduce carbon emissions, this industry is still one of the biggest sources of emissions in the world. Although various technologies have been developed in recent years, there are still serious challenges in this way. Of course, the role of carbon management strategies and standards cannot be ignored easily. These strategies, which include the establishment of laws, development of standards, and executive instructions, have made gas-producing countries and companies more sensitive to their activities. Following these laws and implementing global and regional standards that are currently mandatory in some parts of the world will reduce carbon emissions. However, it should not be overlooked that companies need a series of executive measures in the field of production to implement these standards and comply with the laws. This is why many advanced and innovative technologies have been introduced to the market in recent years, although many of these technologies are in their infancy and need further studies in terms of ease of use, costs of implementation, and compatibility with the industry.

But what attracts attention the most is the decisive role of governments and international organizations in supporting these creative technologies and innovations. According to the income they have, governments alone or by allocating financial resources to companies, universities, and other non-governmental organizations can play a decisive role in the development and expansion of the use of advanced technologies and the implementation of environmental policies and carbon management standards. In addition, the adoption of supportive laws, such as tax exemptions for manufacturers that move towards reducing carbon emissions and environmental impacts, can provide these industries with the incentive and competitiveness to progress further, while these supports make the use of new technologies more cost-effective and their expansion and development faster.

Also, governments and international and regional organizations can play a decisive role in this way by establishing laws and setting goals and obligations for their members. However, this point should not be overlooked that all these material and spiritual support, approval, and implementation of international and regional standards, laws, and agreements should all be done with the aim of simplicity in implementation and cost-effectiveness.

To summarize, the natural gas industry's commitment to reducing methane emissions and managing carbon effectively is evident in the significant technological advancements, regulatory frameworks, collaborative initiatives, and strategic tactical plans underway. These efforts not only mitigate environmental impacts but also enhance the economic sustainability of natural gas production. Continued innovation, supported by robust regulatory frameworks and collaborative partnerships, will be essential in achieving long-term climate goals and ensuring a sustainable energy future. The continuous evolution of these strategies underscores the industry's commitment to reducing its environmental footprint and contributing to global emission reduction goals. The path forward requires sustained innovation, supportive policies, more investment, and robust partnerships to achieve a sustainable energy future.

Author details

Parinaz Soleimani* and Ehsan Hakimi
South Pars Gas Complex (SPGC), Asaluyeh, Bushehr, Iran

*Address all correspondence to: parinaz2005@hotmail.com

IntechOpen

© 2024 The Author(s). Licensee IntechOpen. This chapter is distributed under the terms of the Creative Commons Attribution License (http://creativecommons.org/licenses/by/4.0), which permits unrestricted use, distribution, and reproduction in any medium, provided the original work is properly cited.

References

[1] IEA. World Energy Outlook 2023. France: International Energy Agency; 2023

[2] IEA. Global Methane Tracker 2023. Paris: IEA; 2023. Available from: https://www.iea.org/reports/global-methane-tracker-2023

[3] Sakata S, Zeleke Aklilu A, Rodrigo P. Greenhouse gas emissions data: Concepts and data availability. In: OECD Statistics Working Papers 2024/03. USA: OECD Organisation for Economic Co-operation and Development; 2023

[4] Saunois M, Bousquet P, Poulter B, Peregon A. The global methane budget 2000 to 2012. Earth System Science Data. 2016;**8**(2):697-751

[5] United States Department of Energy. Flaring and Venting Reduction Research & Development Activities. Washington, DC: United States Department of Energy; 2021

[6] IOGP, ipieca, and GGFR. Flaring management guidance for the oil and gas industry. London: IPIECA-IOGP-GGFR; 2021

[7] Lorenzato G, Tordo S, Van Den Berg E, Howells HM, Sarmiento-Saher S. Financing Solutions to Reduce Natural Gas Flaring and Methane Emissions. Washington, DC, USA: The World Bank; 2022

[8] Speight JG. Handbook of Industrial Hydrocarbon Processes. Washington, DC: Elsevier Inc.; 2020

[9] Brancaccio E. GTL: Small Scale and Modular Technologies for Gas to Liquid Industry. Rome, Italy: Oil & Gas portal, no. Serintel Srl; 2018

[10] Subramaniam K, Yao Wong K, Wong KH, Chong CT, Ng J-H. Enhancing biodiesel production: A review of microchannel reactor technologies. Energies. 2024;**17**(1652)

[11] AIChE. The Research Coordination Network (RCN) on Carbon Capture, Utilization and Storage (CCUS) is a Project Funded by the National Science Foundation (NSF) under the Science, Engineering and Education for Sustainability (SEES) [Online]. Available from: https://www.aiche.org/ccusnetwork/what-ccus#:~:text=Share,safe%20and%20permanent%20storage%20options

[12] IEA. The Oil and Gas Industry in Energy Transitions, Insights from IEA analysis. Washington, DC: IEA; 2020

[13] Figueroa J, Fout TE, Plasynski SI. Advances in CO_2 capture technology—The U.S. Department of Energy's Carbon Sequestration Program. International Journal of Greenhouse Gas Control. 2008;**2**(1):9-20

[14] The European Commission. EU ETS Handbook. Belgium, Brussel: The European Commission; 2015

[15] U.S. Methane Emission Reduction Action Plan. The White House Office of Domestic Climate Policy. 2021

[16] IEA. [Online]. 2024. Available from: https://www.iea.org/energy-system/carbon-capture-utilisation-and-storage/co2-capture-and-utilisation

Chapter 3

Simulation and Parametric Sensitivity Study of Recovered Waste Heat for Gas Processing Carbon Capture Unit

Tianci Li, Congning Yang, Teerawat Sema, Christine Chan and Paitoon Tontiwachwuthikul

Abstract

This study developed the presented process simulation based on the design parameters of typical gas plant CCS units in North America. Since the recovered waste heat technology is a new commercial design in solvent regeneration processes, the parametric sensitivity study could provide several detailed optimization methods for further operations. The parametric analysis was conducted by varying MEA concentration (5–7.5 M), the inlet temperature of rich solvent (105–110°C) and reboiler temperature (122–127°C). Additionally, this study developed two split-flow process configurations based on the existing columns and operational conditions. The application of a split-flow configuration with the proper solvent concentration has the potential to save up to 40% of energy costs. The work also evaluated the factors that influence overall capture performance such as column size, type of amine and solvent flow rate. It is promising that the split-flow process configuration and solvent concentration are two potential optimization routes for further practical operations.

Keywords: carbon capture, gas processing, amine sweetening, energy consumption, waste heat recovery

1. Introduction

Post-combustion carbon capture (PCC) is a typical technology for capturing carbon dioxide (CO_2) from flue gases. Throughout the world, PCC will play a crucial role in climate change mitigation in the next few decades since several commercial-scale facilities have operated successfully, such as Boundary Dam Integrated CCS Demonstration and PetraNova Project [1]. PCC is a feasible technology for most existing industrial plants and new facilities as the CO_2 capture unit can be added to existing combustion facilities without significant changes to the existing processes. The technology aims to separate CO_2 from exhaust gases, and the capture process can be achieved using various technologies, including absorption-based process, adsorption-based separation, and membrane separation.

The chemical absorption process is a practical technology that aims to absorb CO_2 using a suitable chemical solvent. Operators can generally select the most appropriate solvent based on the facility's operating conditions and flue gas compositions. Amine-based solvents are the main candidates for commercial applications due to their affordability and excellent capture performance. The selection of amine should rely heavily on the solvent properties and operating conditions of the facility. For instance, primary amines (such as monoethanolamine, MEA) and secondary amines are more reactive with CO_2 due to the direct formation of carbamate by the Zwitterion mechanism [2, 3]. However, tertiary amines (such as N-butyldiethanolamine BDEA, and N-methyldiethanolamine MDEA) exhibit higher absorption capacity and require less regeneration energy for high-pressure operations [4, 5].

Gas power plants are set to play a vital role in the energy transition of Canada's western provinces. In general, natural gas-fuelled power plants have CO_2 streams with lower CO_2 concentrations in the flue gas [6], which means that only active chemical solvents can react well with CO_2. Unfortunately, most active amine solvents, such as primary amines, exhibit high affinity with CO_2, and their regeneration energy is also high. This is primarily caused by the desorption heat of the amines and the sensible heat of the co-solvent, water [7]. It is nearly impossible to altogether remove the water from the amine solutions due to three major reasons: (1) water plays a key role in the chemical reactions of amine solutions and CO_2, the formation of carbamate and carbamate hydrolysis would not occur without water; (2) the effect of the CO_2 solubility; and (3) lowering the boiling point of the liquid solvent, the vapour generation from the reboiler can heat the thermal stripper at a proper temperature range with minimum amine loss to release more CO_2 from the rich solvent.

Researchers have addressed that substituting water for organic solvents can improve the physical solubility of CO_2 [8]. Later on, researchers developed several novel application routes of non-aqueous amine solutions, and they found that blending organic solvent with amines may cause different levels of liquid–liquid phase separation phenomena, known as the bi-phase phenomena. Despite the rich solvent phase exhibiting an impressive high CO_2 absorption capacity, the fluidity issue was also getting serious. When using some novel synthesized amines or polyamines, issues such as high raw material costs, supply shortages, high toxicity, and high volatility can also arise. At the current practical application stage, aqueous amine solutions are still deemed as the key feasible route. Therefore, this work paid attention to common amines, and the process optimization with novel techniques plays a critical role in the solvent regeneration process. Reducing energy penalty can eventually make the process more affordable and optimize the OPEX (Operational Expenditure), making it more feasible [9].

This study aims to optimize the novel waste heat recovery process to improve its overall efficiency. To achieve this goal, several sensitivity analyses were performed by varying the operating variables, such as the temperature of the feeding solvent and reboiler, the solvent flow rate, the amine concentration, and the type of amines.

2. Methodology

This work used the process simulator ProMax® 5.0 with Mass + Heat transfer model to simulate the performance of various process configurations. The reference case for the simulation study was based on practical design data from a Canadian gas plant CCS unit. The summary of the flue gas stream is listed in **Table 1**, and the design data of the columns and equipment can be found in **Table 2**.

ProMax® 5.0 includes several thermodynamic fluid property packages for CO_2 capture process. Amine Sweetening – PR is applied to the acid gas removal process and involves a vapour package of the Peng-Robinson equation of state and a liquid package of Electrolytic. The property package selection depends on the components of the fluids used in the system. In this work, the liquid system was considered a non-ideal system since aqueous amines are polar molecules. Equation of State (EoS) models were employed for gases, as they are Pressure-volume-temperature (PVT) models that mathematically express the phase behaviour of hydrocarbon fluids as a function of pressure and temperature. These property packages allowed for the automatic management of the reactions of aqueous MEA-CO_2 system and the physicochemical properties of the solvent components in the system.

Name	Mol fraction
Nitrogen dioxide	0.00006
Nitrogen	0.74942
Oxygen	0.11120
Carbon Dioxide	0.05591
Carbon monoxide	0.00046
Water	0.07258
Methane	0.00131
Ethane	0.00002
Formaldehyde	0.00009
Benzene	0.00004
Argon	0.00891

Table 1.
Typical exhaust gas composition.

How	Data
Absorber tower size	1678 mm (diameter)|3 (sections) x 5400 mm (height)
Amine flow rate	40.43 m^3/hr
Rich/lean amine pump power	15HP @ 3600 RPM
Differential pressure of amine pump	400 kPag
Desorber (stripper) tower size (Base case)	914 mm (diameter)|2 (sections) x 4000 mm (height)
Reflux pump power	1/2HP @ 1800 RPM
Differential pressure of reflux pump	50 kPa
Gas flow rate	686 E^3 m^3/day

Table 2.
Design data of columns and other mechanical equipment.

In this work, process's overall performance can be evaluated using three performance indicators: cyclic capacity of the solvent, reboiler duty, and overall CO_2 capture efficiency. The cyclic capacity of the solvent refers to the difference between the lean loading (α_{lean}) and rich loading (α_{rich}), where the CO_2 loading (α) is represented as the ratio of the number of moles of CO_2 to the number of moles of amine in the liquid phase. The reboiler duty, also known as the regeneration energy, is the sum of the desorption heat, sensible heat, and heat of vaporization. The overall CO_2 capture efficiency is a macroscopic indicator expressed as the percentage ratio of mass concentration of the CO_2 in the inlet and outlet gas streams. This work evaluated the simulation result validation by using the percentage of average deviation (%AAD), which can be expressed as the equation below:

$$\%\text{AAD} = \frac{1}{\text{n}}\Sigma\frac{\text{SIM - REF}}{\text{REF}} \times 100 \tag{1}$$

where %AAD is the average absolute deviation; SIM and REF represent simulation results obtained from the simulation (this work) and the original onsite data, respectively; n is the number of data.

3. Process description and simulation validation

3.1 Conventional process with waste heat recovery process

As shown in **Figure 1**, the waste heat recovery process involves replacing traditional steam regeneration with energy released from the hot flue gas stream. The names of the equipment and streams on the process flowsheet are explained in **Table 3**. The energy conversion primarily depends on the required heat transfer rate, which is affected by the size of the heat exchanger. In ProMax simulation, if the user cannot attach the heat exchanger's surface area to the system, they must input the inlet and outlet temperature of the heating medium. In this work, the ProMax system calculated the converted energy (stream Q-2) based on the outlet stream temperature (stream 14) with an assumption of 30% heat loss. Due to the energy conversion, the temperature of the Flue Gas stream can be cooled from 449–180°C, equivalent to 2033.38 kW of recovered heat. Compared to the onsite data, this represents a 3.5% of %AAD. A flue gas cooler (XCHG-101) is used to cool the flue gas from 180–38°C, equivalent to 1113.98 kW of energy consumption, to provide additional cooling energy. Compared to the onsite data, this represents an 11.8% of %AAD. In addition to the flue gas cooler, the amine cooler and reflux condenser require significant amounts of power, while the power required by the amine pumps is relatively tiny. The simulated power consumption was consistent with onsite data within an acceptable deviation, indicating the reliability and value of the further predictions.

3.2 Split-flow process configurations with waste heat recovery process

The idea of using split-flow design in carbon capture is to improve the overall efficiency and cost-effectiveness of the process. Typically, solvent Split flow is a promising strategy that aims to reduce the solvent loaded in the heat exchanger and eventually improve the thermal transfer rate. Sometimes, the solvent streams are split and recirculated in the absorber or desorber to maximize the CO_2 absorption capacity [1]. Although the absorption/desorption kinetics can be improved,

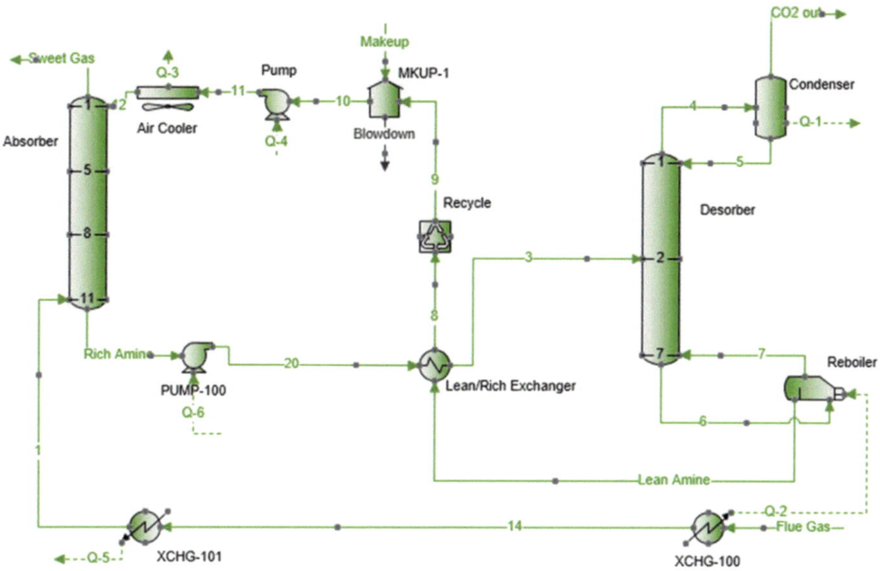

Figure 1.
Conventional process flowsheet with waste heat recovery process.

Name of the icon	Name of the equipment/stream
Pump-100	Rich amine pump
Pump	Lean amine pump
XCHG-100	Waste heat exchanger
XCHG-101	Flue gas cooler
MKUP-1	Make-up tank
Stream 1	Flue gas inlet
Stream 2	Lean solvent inlet
Stream 3	Hot-rich solvent feeds
Stream 4	Separated wet gas
Stream 5	Condensed water
Stream 6	Treated solvent from desorber
Stream 7	Hot vapour
Stream 8	Warm lean solvent

Table 3.
Important naming of the equipment and streams on the simulation flowsheet.

the main limitation of this method is that the overall performance dramatically relies on the type of solvent. Some active amine solvents (such as primary amine solutions) have insufficient absorption capacity is the main limitation, regardless of the CO_2 partial pressure conditions. This work aims to develop novel split-flow process configurations integrated with waste heat exchangers. Consequently, this study will find different new promising routes for waste heat recovery processes for the gas CCS plant.

As mentioned earlier, the sensible heat of the solvent is primarily determined by the water content of the solvent. By using a split-flow design, the temperature distribution of the fluid can be optimized, resulting in a reduction in reboiler duty. Besides, the thermal transfer efficiency in the lean/rich exchanger would be more beneficial due to the lower solvent flow rate. As shown in **Figure 2(a)** and **(b)**, there are three major new equipment elements in the split-flow processes: a solvent splitter (SPLT-101) and two heating elements (XCHG-104 and XCHG-102). XCHG-102 is a continuous waste heat recovery unit designed to transfer heat from the hot flue gas to the heating medium. This heated medium then serves as the heating source for warming the rich solvent in SCHG-104. The split stream of rich solvent flows into the secondary waste heat exchanger and then to the desorber (Stream 18). With this

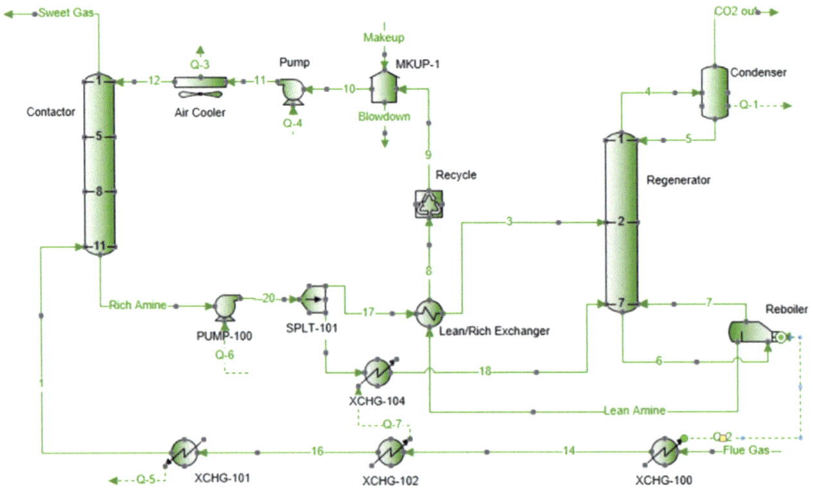

(a) Split flow #1

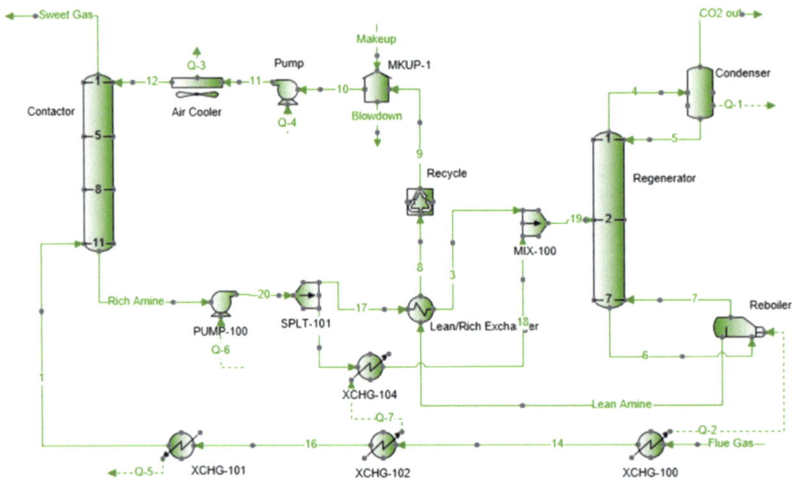

(b) Split flow #2

Figure 2.
Split-flow design of thermal desorber integrated with waste heat exchange process.

design, the duty of the flue gas cooler would be reduced while the temperature of the rich solvent feeds would be improved.

Compared to two split-flow process configurations, **Figure 2(a)** aims to directly split the solvent to the bottom of the desorber. Due to the water vaporization phenomena, Stream 18 will provide more heat to the entire desorption column. Additionally, the desorber is a thin column, and a lower liquid flow rate passing through the structure packing can potentially reduce the risk of flooding. This design will be beneficial for further practical operations if operators want to try other viscous chemical solvents. By reducing the feed flow rate, the residence time of the liquid in the desorber increases. This allows more time for CO_2 to be released from the solvent, improving the overall efficiency of the desorption. A lower flow rate can improve the mass transfer between the liquid and gas phases. From the perspective of operational flexibility, reducing the flow rate can allow the process to better adapt to variable CO_2 loads, optimizing the mass transfer performance by providing the lowest need for hot steam. **Figure 2(b)** aims to develop a secondary heat transfer process for the rich solvent feed (Stream 19) to eventually reduce boiler duty.

4. Results and discussion

4.1 The effect of MEA concentration on the overall performance

This study investigated the effect of MEA concentration on the overall performance. The conventional process flowsheet was utilized and integrated with the waste heat recovery process for the simulation of this section. In addition, the temperature of the rich solvent feed was varied from 105–110°C. The temperature range is reasonable when compared to practical operations at the CCS plant.

Table 4 shows that the regeneration energy is reduced with increasing the concentration of MEA, which can be attributed to the lower solvent flow rate demand and higher CO_2 loading. A solvent with higher CO_2 loading is more favourable in the regeneration process due to the larger driving force of CO_2, which is beneficial for mass transfer phenomena. However, this study also found that the lean loading could

Solvent	Overall CO_2 capture efficiency, %	Solvent flow rate, m^3/hr	Regeneration energy, GJ/tCO_2	Lean loading-rich loading
5 M MEA	77.37	35	4.44	0.25–0.47
	77.76	40	4.40	0.28–0.47
	78.19	45	4.42	0.29–0.47
6.25 M MEA	80.98	35	4.23	0.30–0.48
	81.44	40	4.21	0.31–0.48
	81.22	45	4.21	0.32–0.48
7.5 M MEA	82.43	30	4.16	0.30–0.49
	84.21	35	4.10	0.32–0.48
	83.93	40	4.45	0.34–0.48

Table 4.
The effect of MEA concentration on the overall performance ($T_{rich,feed}$ = 105°C; T_{lean} = 122°C).

be increased by increasing the concentration of MEA, possibly due to the consistent energy conversion from the waste heat exchanger. A similar tendency of regeneration performance was observed in **Table 5**.

From **Tables 4** and **5**, it can be observed that achieving 85% overall CO_2 capture efficiency is difficult when the temperature of the rich solvent feed is lower than 110°C. There are two simplest methods that can be considered to resolve the problem: (1) lower the lean loading; and (2) increase the solvent flow rate. However, increasing the solvent flow rate at the CCS plant is challenging, as the desorber is a thin thermal tower, and an excessive solvent flow rate could result in flooding. Therefore, the first option of lowering the lean loading is more feasible. Two strategies to achieve this target are: (1) constructing new heat suppliers to lower the lean loading; and (2) increasing the rich solvent temperature.

For general practical operations, it is not economically feasible to keep adding new thermal generators to heat the lean solvent to become leaner. Additionally, since the desorber column has already been built, replacing it with a taller one is nearly impossible. Therefore, increasing the rich solvent temperature is deemed as a relatively feasible solution to lower the lean loading of solvent. Additionally, the enhancement behaviour of a higher temperature of the rich solvent feed could also be affected by the concentration of amine. **Tables 4** and **5** show that the regeneration energy slightly increases by increasing the concentration of MEA from 6.25 M to 7.5 M, although the overall CO_2 capture efficiency is still growing. The higher lean loading and unexpectedly low residence time at high solvent flow rate conditions possibly cause this. Besides, the physical properties of the solvent, such as viscosity and density, also affect the solvent flow rate. Therefore, determining the optimal concentration and flow rate of the solvent requires finding a balance between overall CO_2 capture efficiency, energy consumption and optimal residence time.

The flooding fraction is a function of fluid flow rate, which significantly affects the CO_2 loading in the solvent. When the solvent flow rate exceeds the absorber's capacity, which is determined by the diameter of the column, flooding issues occur, resulting in reduced efficiency and increased energy consumption due to lower rich loading. **Table 5** shows that by increasing the MEA concentration from 5 M to 6.25 M, the overall CO_2 capture efficiency can be maintained at around 85% while

	Overall CO_2 capture efficiency, %	**Solvent flow rate, m^3/hr**	**Regeneration energy, GJ/tCO_2**	**lean loading-rich loading**
5 M MEA	78.53	35	4.37	0.25–0.47
	82.24	40	4.40	0.25–0.47
	84.68	45	4.20	0.28–0.47
6.25 M MEA	79.77	30	4.20	0.27–0.46
	84.07	35	4.18	0.28–0.47
	85.51	40	4.12	0.29–0.47
7.5 M MEA	80.42	25	4.27	0.28–0.50
	85.66	30	4.10	0.30–0.50
	88.28	40	4.15	0.32–0.49

Table 5.
The effect of MEA concentration on the overall performance ($T_{rich,feed}$ = 110°C; T_{lean} = 122°C).

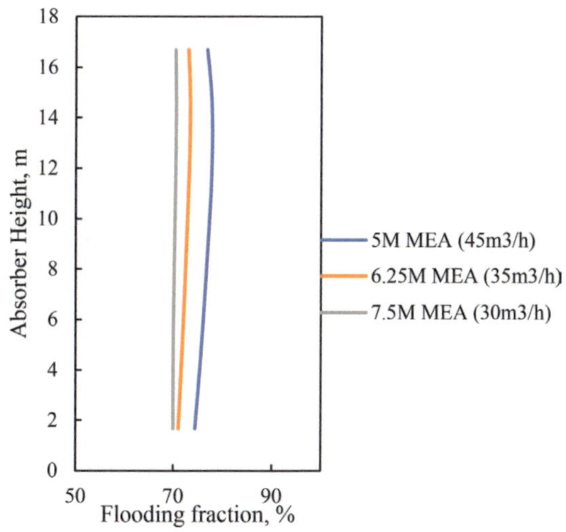

Figure 3.
The effect of MEA concentration on the absorber flooding fraction.

reducing the solvent flow rate by at least 10%. **Figure 3** compares flooding fraction in the absorber based on maintaining the overall CO_2 capture efficiency around 85% (±1%). Generally, an appropriate flooding fraction range is between 50 and 75% in the absorber. To ensure reliable predictions, ProMax warns users once the flooding fraction is more significant than 80%, the system does not accept the input if the flooding fraction is higher than 100%. Properly increasing solvent concentration can achieve around a 5% reduction in flooding fraction caused by the reduced flow rate. It should be addressed that the flooding fraction is also determined by the ProMax system based on the specific process and operating conditions, including the gas flow rate, amine solution properties and internal geometry of the column.

4.2 The effect of lean loading on the overall performance

As discussed in the previous section, the solvent flow rate is an important factor affecting the regeneration energy required to release the absorbed CO_2 from the liquid phase. Increasing the solvent flow rate can increase the pressure drop in the desorber, which in turn can increase the regeneration energy. On the other hand, an insufficient solvent flow rate can also increase the regeneration energy, as shown in **Table 6**. This is because the CO_2 lean loading is reduced by increasing the temperature of the reboiler. **Table 6** can be divided into two parts: (1) for 5 M MEA operations, the optimal solvent flow rate is between 40 and 45 m^3/hr.; (2) for 6.25 M MEA operations, the optimal solvent flow rate is between 35 and 40 m^3/hr. It is noteworthy that increasing the lean solvent temperature from 122–127°C can significantly improve the overall CO_2 capture efficiency. Generally, 127°C is considered a safe high temperature with an acceptable thermal degradation rate for MEA operations [10].

However, the reality poses a challenge. When developing a new process for a practical CCS plant, it is crucial to consider both the economic and technical feasibility. In terms of the technical aspect, one potential strategy to address the regeneration energy issue is to utilize a waste heat exchanger to harness the waste heat from the flue gas source. However, the flue gas flow rate is assumed to be constant, and it cannot be directly used to heat the rich solvent because of its high temperature, which can cause

Solvent	Overall CO_2 capture efficiency, %	Solvent flow rate, m^3/hr	Regeneration energy, GJ/tCO_2	Lean loading-rich loading
5 M MEA	81.56	28	4.43	0.18–0.47
	83.31	31	4.28	0.20–0.47
	83.18	32	4.08	0.20–0.47
	83.03	35	3.97	0.22–0.47
	81.91	40	3.88	0.23–0.47
	81.77	45	3.82	0.25–0.47
6.25 M MEA	86.98	25	4.21	0.21–0.48
	88.19	29	4.02	0.23–0.48
	88.73	32	3.83	0.25–0.48
	88.71	35	3.75	0.26–0.48
	88.44	40	3.72	0.28–0.48
	88.22	45	3.74	0.29–0.48

Table 6.
The overall performance when increasing reboiler temperature ($T_{rich,feed}$ = 105°C; T_{lean} = 127°C).

thermal degradation issues. This implies that an additional operational unit of waste heat exchanger should be employed. Alternatively, a traditional thermal transfer reboiler with a heat generator must be applied.

4.3 Split-flow process

As mentioned in the previous sections, increasing the inlet temperature of the rich solvent is a relatively feasible way to reduce the lean CO_2 loading of the solvent because it does not need a heat generator. However, there are two major limitations: (1) an additional cross-heat exchanger is required, and (2) limited heat available from hot lean solvent. These two limitations make it challenging to increase the inlet temperature of the rich solvent to higher than 110°C. Therefore, the technique of split-flow process can certainly address these issues.

Additionally, the split-flow process can increase the temperature of lean solvent. By heating the split solvent stream to as high as 127°C, then the lean loading of the solvent is reduced, improving the cyclic capacity.

4.3.1 The overall performance

For split-flow process configurations, it is essential to determine the appropriate split ratio of the solvent. In this work, the size of lean/rich cross-heat exchanger is held constant, and the maximum temperature lean solvent is fixed at 122°C. This implies that if the flow rate of the solvent passing through the cross-heat exchanger is too low, then the outlet lean amine temperature will be high, increasing the duty of amine cooler. On the other hand, an excessive solvent flow rate may compromise thermal transfer performance and negatively impact the overall process efficiency.

Figure 4 shows the temperature profile of the outlet lean solvent from the desorber. In this profile, the temperature of lean amine stream is also higher due to the higher temperature of the outlet lean solvent (stream 6). It can observed that

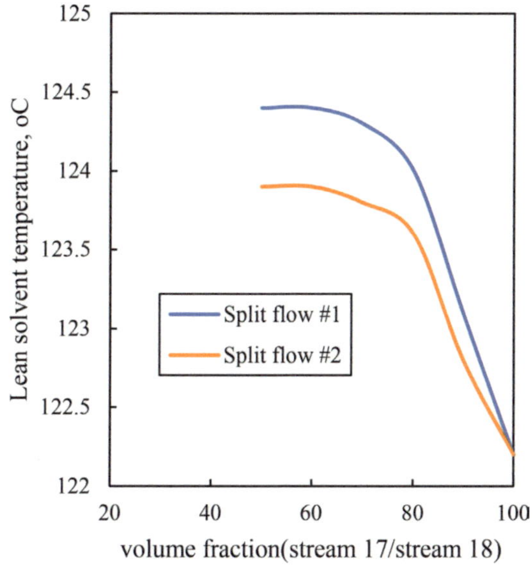

Figure 4.
Temperature profile of 5 M MEA split-flow process configurations.

splitting 20% of the rich solvent to the secondary waste heat exchanger is within the optimal range because the temperature gradient becomes flat once the split ratio exceeds 80%. In this case, this work specifies a split ratio of 20% for all predictions.

Table 7 presents a comparison of the two split-flow process configurations. The comparison highlights the dependency of the process performance on the concentration of MEA solvent. Split flow #1 is more beneficial for operations with a higher concentration solvent due to two primary possible reasons: the physical properties of the solvent and the solvent flow rate. In fact, both parameters affect the residence time of the fluid in the desorber column, which is the main factor of pressure drop. The process with higher concentration MEA requires a lower solvent flow rate, but the fluidity issue becomes more severe due to the higher viscosity, as shown in **Figure 5**.

Solvent	Overall CO_2 capture efficiency, %	Solvent flow rate, m^3/hr	Regeneration energy, GJ/tCO_2	Lean loading-rich loading
		Split flow #1		
5 M MEA (Benchmark)	80.72	42	4.38	0.27–0.47
6.25 M MEA	85.32	39	4.11	0.30–0.48
7.5 M MEA	87.13	34	4.01	0.31–0.48
		Split flow #2		
5 M MEA (Benchmark)	80.93	42	4.38	0.27–0.47
6.25 M MEA	85.32	39	4.14	0.30–0.48
7.5 M MEA	86.88	34	4.06	0.31–0.48

Table 7.
Comparison of two Split flow process configurations.

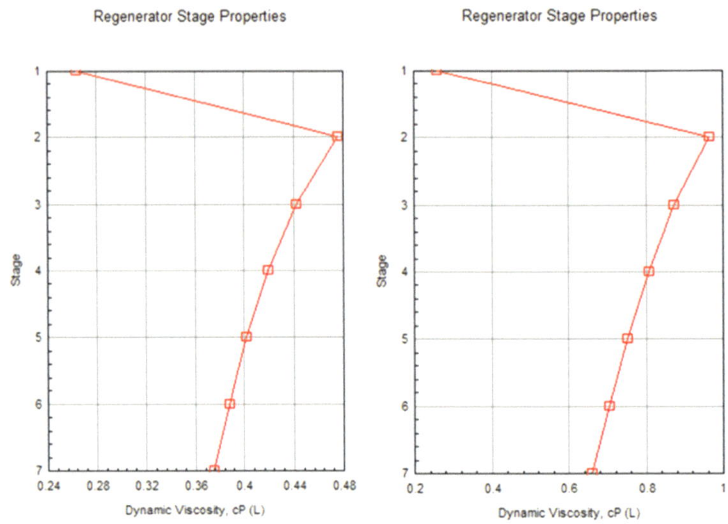

(a) 5M MEA (b) 7.5M MEA

Figure 5.
Predictions of solvent viscosity profile from ProMax (Split flow #1).

Solvent with higher viscosity poses challenges for thermal transfer in the reboiler and may cause a higher pressure drop due to the longer residence time in the column. For Split flow #1, less solvent passes through the packing, which can partially overcome the long residence time issue. Compared to the conventional process, the fluidity issue is more effectively relieved due to the higher inlet temperature of the rich solvent. On the other hand, Split flow #2 is a good design for solvents with lower amine concentrations due to the more excellent contact of the hot vapour and solvent in the desorber column.

4.3.2 The effect of solvent flow rate on the pressure drop

The pressure drop in the desorber column is greatly affected by the hydrodynamic characteristics, which are determined by several factors, including vapour/solvent flow rate ratio, temperature, solvent properties and the column size [11]. An optimal pressure drop provides an optimal driving force for mass transfer between the solvent and gas phase, which promotes the release of CO_2 from the liquid phase. However, excessive high-pressure drops can lead to flooding and other safety-related operational issues, reducing efficiency and increasing operational risks.

In general, stripping pressure can significantly affect the amine regeneration performance in terms of regeneration energy and efficiency. In this work, the flow gas pressure at the bottom section of the desorber column is specified as 200 kPa. **Figure 6** illustrates the impact of solvent flow rate on the pressure drop for different concentrations of MEA operations. Due to the higher inlet temperature of the rich solvent feed, Split flow #2 can successfully reduce the pressure drop compared to the conventional process. For Split flow #1, the pressure drop reduction is more significant due to the column's higher gas/liquid ratio, making it more practical for concentrated solvent operations.

At high solvent flow rate conditions, split-flow processes exhibited a smaller pressure drop which ultimately reduces the energy required by the reboiler as it helps

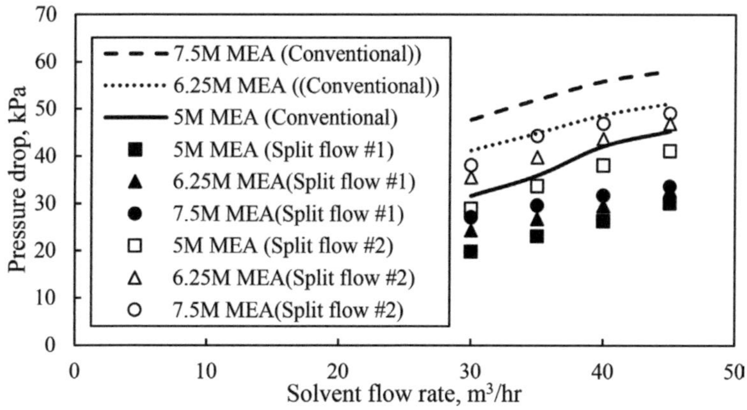

Figure 6.
Comparison of pressure drop at a varied flow rate.

maintain a more consistent temperature profile along the desorber, which is critical for efficient stripping. Smaller pressure drops decrease the risk of flooding, leading to more stable and reliable operation of the desorber.

4.4 The effect of the absorber column on the overall CO_2 capture efficiency and regeneration energy

When designing a CO_2 absorber column using theoretical calculation methods, the height and diameter of the column are influenced by different factors related to its performance and operational efficiency. The required CO_2 capture efficiency primarily determines the height of the CO_2 absorber column. This is because the height needs to provide sufficient contact time and surface area for the gas and liquid phases to interact, allowing the CO_2 to be absorbed effectively by the solvent. Theoretically, the height is often calculated based on the number of equilibrium stages needed to achieve the desired capture efficiency, its influencing factors include CO_2 concentration, solvent type and concentration, and the desired purity of the treated gas. While the fluid flow rates of the gas and liquid phase influence the diameter of the absorber column. The flooding factor plays an essential role, in measuring the maximum gas and liquid flow rates the column can handle before flooding occurs. Flooding issue occurs when the gas flow is too high, causing the liquid to be entrained by the gas flow, which indirectly leads to a loss of separation efficiency. In this work, as shown in **Table 2**, the base case scenario demonstrates that when the diameter of the absorber is set to 1678 mm, it can handle a maximum of 686 E3 m^3/day of gas flow.

As shown in **Table 8**, increasing the column diameter while maintaining the same inlet gas flow rate can reduce regeneration energy and increase overall CO_2 capture efficiency. From this work, it is observed that changing the column diameter reduces the flooding issue and indirectly improves the solvent's CO_2 absorption capacity.

4.5 The effect of amine on the overall CO_2 capture efficiency and regeneration energy

According to previous publications [12, 13], for less reactive amine solvents, blending piperazine is a good strategy to increase the reaction kinetics with CO_2. Additionally, an appropriate volume of piperazine (PZ) can improve the solubility of

Diameter of absorber, mm	Overall CO_2 capture efficiency, %	Regeneration energy, GJ/ tCO_2	Lean loading-rich loading
Split flow #1			
1678	85.32	4.11	0.30–0.48
2678	86.12	3.99	0.28–0.50
3198	88.13	3.89	0.28–0.51
Split flow #2			
1678	85.32	4.14	0.30–0.48
2678	86.12	4.01	0.28–0.50
3198	88.13	3.91	0.28–0.51

Table 8.
Comparison of different diameters of the columns using 6.25 M MEA in two Split flow process configurations (Solvent flow rate = 39 m^3/hr).

CO_2 in the liquid phase. However, the use of piperazine may cause solid formation in cold regions leading to potential operation risks, and the possibility of solid formation is increased by higher the piperazine concentration. Therefore, this work uses only 1 M PZ to facilitate the amine solvent.

As shown in **Table 9**, maintaining the amine concentration consistently at 5 M, substituting MEA with methyldiethanolamine (MDEA) without the aid of piperazine, reduces the overall CO_2 capture efficiency, even though the regeneration energy is attractive. When MDEA is substituted by 2-amino-2-methyl-1-propanol (AMP), the overall CO_2 capture efficiency significantly increases, but the regeneration energy also increases. The tri-solvent candidate 2 M MDEA +1 M PZ + 2 M MEA performs the best in terms of overall CO_2 capture efficiency, regeneration energy, and absorption capacity. However, it should noted that this tri-solvent is much complex than single MEA and other solvent blends, leading to several potential challenges for

Solvent	Overall CO_2 capture efficiency, %	Regeneration energy, GJ/ tCO_2	Lean loading-rich loading
Split flow #1			
5 M MEA	80.72	4.11	0.27–0.47
2 M MDEA+3 M MEA	77.98	3.26	0.22–0.41
2 M AMP + 3 M MEA	85.11	3.69	0.26–0.48
2 M MDEA +1 M PZ + 2 M MEA	90.17	3.29	0.30–0.54
Split flow #2			
5 M MEA	80.93	4.11	0.27–0.47
2 M MDEA+3 M MEA	77.98	3.24	0.22–0.41
2 M AMP + 3 M MEA	85.12	3.68	0.26–0.48
2 M MDEA +1 M PZ + 2 M MEA	90.17	3.27	0.30–0.54

Table 9.
Comparison of different 5 M amines in two split-flow process configurations (Solvent flow rate = 42 m^3/hr).

practical operations. These challenges include maintaining mass balance with make-up solvent, degradation issues due to more diverse products formed from the reaction, and potential solid formation. Once the operators address these challenges, using the tri-solvent could be a good strategy.

By comparing the two split-flow configurations, the overall performances are very close. From the simulation results, the most sensitive factor appears to be temperature. Indeed. thermal behaviour in both the absorption and desorption processes is important, as it impacts not only the kinetics but also the absorption capacity.

4.6 Variable operating cost analysis

The Variable Operating Cost (VOC) mainly includes the costs of solvent and power consumption during the operation. It is an essential indicator in estimating the total annual cost of the process. The other two indicators are the fixed operating costs and annual capital expenditures. In this study, the VOC analysis shows the influence of amine concentration and Split flow process from an economic aspect for practical operations.

Traditionally, the solvent regeneration process accounted for at least 60% of the total energy demand in the traditional CO_2 capture process without considering the work done from the desorbed CO_2 compression unit [2, 12]. Among all energy consumers in the solvent regeneration process, reboiler is the primary sector. The heat required for the reboiler is traditionally provided by the steam regenerator. However, with novel waste heat recovery processes, the traditional heat extraction method can be partially or completely eliminated. The remaining energy consumers include pumps and coolers for liquid and gas, respectively. **Figure 7(a)–(c)** shows a detailed

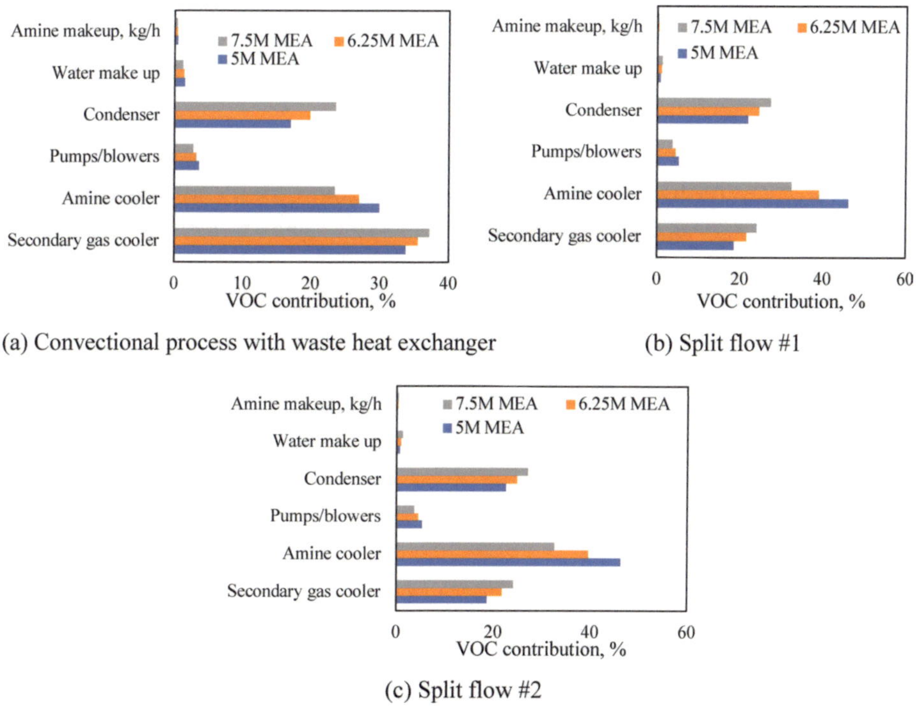

Figure 7.
Comparative variable operating costs of various process configuration.

distribution diagram of the VOC contribution. With the application of the secondary waste heat recovery system, the duty of flue gas cooler is reduced while the condensation power and amine cooler are slightly increased. Split flow #1 and Split flow #2 tend to have similar energy contributions.

Figure 8 displays the comparison of VOC savings with the application of different processes and solvents. By comparing with the benchmark case (5 M MEA operation with waste heat recovery process), changing the concentration of the solvent is feasible when considering the overall CO_2 removal efficiency at a constant 85%. Even though the application of split-flow with concentrated amine solvent exhibited slightly more significant savings, their equipment and retrofitting costs are also higher. Two split-flow process configurations exhibit better advantages as the overall CO_2 removal efficiency increases to 90%. This is based on the assumption that increasing the overall CO_2 removal efficiency to 90% requires additional heat supply. Conversely, the optimal concentration of MEA should be 6.25 M MEA. With the aid of split-flow design, 6.25 M MEA can provide greater performance and save more energy costs.

For long-term operations, the performance of a carbon capture facility largely depends on the lifespan of the solvent and its degradation issues, which can be categorized into thermal and oxidative degradation. Current technologies allow operators to extend solvent life by adding additives, which can be either homogeneous or heterogeneous with the solvent. However, in simulation studies, there is a knowledge gap due to the lack of kinetic and thermodynamic data for solvent-additive interactions. This makes it challenging to accurately model the process using the current dataset.

By utilizing a waste heat exchanger in the conventional process, the energy demand in the desorption process shifts from heating to cooling, which is beneficial for operators. Typically, heating elements in solvent regeneration consume over 60% of the total energy in traditional processes. The demand for solvent is closely linked to overall efficiency. According to market surveys, the bulk price of MEA typically ranges from USD 7–15 per kilogram. If a blended solvent is used, the estimated cost increases slightly, as the unit prices of AMP and PZ are generally 3–5 times higher than that of MEA. As discussed in the previous section, while blended solvents can reduce generation energy, the economic benefits of this reduction are limited due to the use of waste heat recovery processes. In this context, the primary option would be to optimize the concentration of MEA. However, it is challenging to conclude that

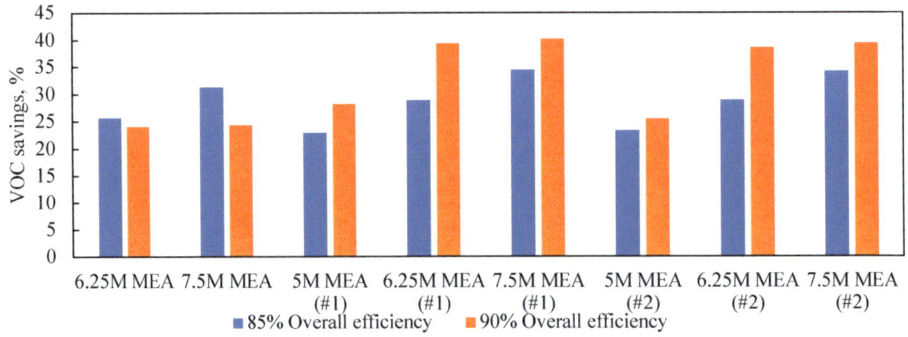

Figure 8.
Percentage VOC savings for various process configuration. (Reference case: 5 M MEA operation with waste heat recovery process).

using blended solvents offers no benefits. PZ and AMP are amines with superior thermal stability, which could enhance overall capture stability in long-term operations.

4.7 Limitations of simulation and potential errors

In developing and optimizing carbon capture processes using ProMax, several limitations and potential sources of error are considered: (1) Simplified assumptions: The simulation relies on thermodynamic models to predict the behaviour of the solvent and other process variables. The simulator can indeed predict potential risks such as flooding issues and mass and heat balances due to vaporization and solvent loss. However, the model may not fully capture the complexities of real-world operations, the operating challenges such as foaming issues, fouling and heat loss are not considered; (2) Data gaps: the simulation cannot fully account for solvent degradation over time, especially for blended solvents. This can lead to overestimation of long-term process efficiency and underestimation of operational challenges; and (3) Equipment modeling limitations: The simulation is typically based on steady-state conditions, which may not capture the dynamic behaviour, making it difficult to model precise stability for 24/7 operations.

5. Conclusions

This is a simulation study based on the integration of a novel waste heat recovery process and a traditional CO_2 capture process located in a gas CCS plant in Canada. This study comprehensively investigates the improvement of several split-flow process configurations while addressing their limitations and advantages. The development of novel process configurations is based on a case of CCS facility including the size of the facility area, pipeline, current equipment setup, and the shape of the columns. The overall performance of the CO_2 capture process is evaluated in terms of energy consumption and CO_2 removal efficiency. The effect of solvent concentration at different temperature and pressure conditions is also assessed. In addition, influencing factors such as the type of solvent and the size of the column are evaluated based on the given flue gas conditions. The results confirm that the selection of solvent concentration and process configuration should be considered based on the operating conditions and targeted overall CO_2 removal efficiency.

Acknowledgements

We gratefully acknowledge the generous support from research grants of the Natural Science and Engineering Research Council of Canada (NSERC), Mitacs Canada and Entropy Inc. In addition, we would like to especially acknowledge the leadership of the late Rick Bower, one of the co-founders of ABC Engineering Inc., who showed us the real need for carbon capture of our time.

Conflict of interest

The authors declare no conflict of interest.

Author details

Tianci Li, Congning Yang, Teerawat Sema, Christine Chan
and Paitoon Tontiwachwuthikul*
University of Regina, Regina, SK, Canada

*Address all correspondence to: paitoon@uregina.ca

IntechOpen

© 2024 The Author(s). Licensee IntechOpen. This chapter is distributed under the terms of the Creative Commons Attribution License (http://creativecommons.org/licenses/by/4.0), which permits unrestricted use, distribution, and reproduction in any medium, provided the original work is properly cited.

References

[1] Li T, Yang C, Tantikhajorngosol P, Sema T, Tontiwachwuthikul P. Studies on advances configurations of post-combustion CO_2 capture process applied to cement plant flue gases. Carbon Capture Science and Technology. 2022;**4**:100064. DOI: 10.1016/j.ccst.2022.100064

[2] Nwaoha C, Beaulieu M, Tontiwachwuthikul P, Gibson M. Techno-economic analysis of CO_2 capture from a 1.2 million MTPA cement plant using AMP-PZ-MEA. International Journal of Greenhouse Gas Control. 2018;**78**:400-412. DOI: 10.1016/j.ijggc.2018.07.015

[3] Shi H, Fu J, Wu Q, Huang M, Jiang L, Cui M, et al. Studies of the coordination effect of DEA-MEA blended amines (within 1+ 4 to 2 +3 M) under heterogeneous catalysis by means of absorption and desorption parameters. Separation and Purification Technology. 2020;**236**:116179. DOI: 10.1016/j.seppur.2019.116179

[4] Li T, Yang C, Tantihajorngosol P, Sema T, Shi H, Tontiwachwuthikul P. Experimental investigations and the modeling approach for CO_2 solubility in aqueous blended amine systems of monoethanolamine, 2-amino-2-methyl-propanol, and 2-(butylamino)ethanol. Environmental Science and Pollution Research. 2022;**29**:s11356. DOI: 10.1007/s11356-022-20411-x

[5] Liu H, Chan V, Tantikhajorngosol P, Li T, Chan C, Tontiwachwuthikul P. Novel machine learning model correlating CO_2 equilibrium solubility in three tertiary amines. Industrial and Engineering Chemistry Research. 2022;**61**:1420-14032. DOI: 10.1021/acs.iecr.2c02006

[6] Kothandaraman A, Nord L, Bolland O, Herzog HJ, McRae J. Comparison of solvents for post-combustion capture of CO_2 by chemical absorption. Energy Procedia. 2009;**1**:1373-1380. DOI: 10.1016/j.egypro.2009.01.180

[7] Oyenekan B, Rochelled G. Alternative stripper configuration for CO_2 capture by aqueous amines. AIChE Journal. 2007;**53**:3144-3154. DOI: 10.1002/aic.11316

[8] Garcia M, Knuutila H, Aronu U, Gu S. Influence of substitution of water by organic solvents in amine solutions on absorption of CO_2. International Journal of Greenhouse Gas Control. 2018;**78**:286-305. DOI: 10.1016/j.ijggc.2018.07.029

[9] Li T, Yang C, Tantikhajorngosol P, Sema T, Liang Z, Tontiwachwuthikul P, et al. Comparative desorption energy consumption of post-combustion CO_2 capture integrated with mechanical vapor recompression technology. Separation and Purification Technology. 2022;**294**:121202. DOI: 10.1016/j.seppur.2022.121202

[10] Muchan P, Saiwan C, Narku-Tetteh J, Idem R, Supap T, Tontiwachwuthikul P. Screening tests of aqueous alkanolamine solutions based on primary, secondary, and tertiary structure for blended aqueous amine solution selection in post-combustion CO_2 capture. Chemical Engineering Science. 2017;**170**:574-582. DOI: 10.1016/j.ces.2017.02.031

[11] Almoslh A, Aghel B, Alobaid F, Heinze C, Epple B. Experimental study of the influence of different load changes in inlet gas and solvent flow rate on CO_2 absorption in a sieve tray column. Entropy. 2022;**24**:1318. DOI: 10.3390/e24091318

[12] Liang Z, Rongwong W, Liu H, Fu K, Gao H, Gao F, et al. Recent progress and new developments in post-combustion carbon-capture technology with amine based solvents. International Journal of Greenhouse Gas Control. 2015;**40**:26-54. DOI: 0.1016/j.ijggc.2015.06.017

[13] Yang C, Li T, Tantikhajorngosol P, Sema T, Xioa M, Tontiwachwuthikul P. Evaluation of novel aqueous piperazine-based physical-chemical solutions as biphasic solvents for CO_2 capture: Initial absorption rate, equilibrium solubility, phase separation and desorption rate. Chemical Engineering Science. 2023;**277**:118852. DOI: 10.1016/j.ces.2023.118852

Section 2

Unconventional Gas Resources: Challenges and Development

Chapter 4

Perspective Chapter: Challenges and Opportunities in Unconventional Gas Resources

Fengyang Xiong, Yiwen (Sophia) Gong and Shu Jiang

Abstract

Unconventional gas resources significantly contribute to the primary production of gas supply worldwide. This chapter delves into the complexities and prospects in key unconventional gas resources, including coalbed methane, tight gas, shale gas, gas hydrates, and hydrogen. It approaches in two directions: (1) theoretical and experimental research and (2) current surveillance methodology, comprehensive modeling and simulation, and production practices. A brief introduction and critical literature review will be summarized on each unconventional gas resource, followed by an analytical discussion, sorted challenges, and potential solutions or opportunities. This chapter attempts to provide academia and industry practical insights on the status, challenges, and future development opportunities of unconventional gas resources.

Keywords: coalbed methane, tight gas, shale gas, gas hydrates, hydrogen

1. Introduction

At the beginning of the twenty-first century, a trend of carbon neutrality (aka, Net Zero) arose worldwide due to a universal concern about pressing global climate changes [1]. Developing and developed countries have been attempting to accelerate the energy transition from conventional fossil fuels toward green sustainable resources (e.g., wind, solar, nuclear, and geothermal energies) [2]. Owing to an insufficient and unstable supply of green sustainable resources resulting from currently high operation costs and lacking key techniques and production strategies, methane and hydrogen were regarded as the bridge from fossil fuels to sustainable resources during this energy transition [3, 4].

In the past decades, methane that is mainly from coalbed methane (CBM), tight gas, shale gas, and gas hydrates have attracted abundant attention from academia and industry [5–12]. Hydrogen is thought as the potential alternative resource that could replace methane and further decrease carbon emissions [13]. To differentiate from conventional fossil fuels (e.g., oil and gas from highly porous or permeable sandstone and carbonate reservoirs), the term "unconventional gas resources" was proposed to refer to gas resources (mostly methane) that are not produced using conventional gas production techniques (e.g., drilling, injection, and production) and require specific techniques for commercial production (e.g., horizontal drilling and hydraulic fracturing) [14].

IntechOpen

Unconventional gas resources were also referred to by other scholars as methane hydrates, biogas, CBM, flue gas, tight gas, shale gas, synthesis gas (a combined gas body of carbon monoxide and hydrogen), biogas, and others [15]. The unconventional resource development ensures the energy supply of the currently growing population and economics and promotes the realization of Net Zero [16]. Based on the U.S. Energy Information Administration's International Energy Outlook 2023, the global primary energy consumption of natural gas in 2025 differs when economic conditions differ [16]. In both low and high scenarios, natural gas and coal exhibit a steady to slightly increasing trend. Our focus on non-fossil fuel sectors, particularly in other cleaner gas resources such as methane hydrates and hydrogen, shows rapid growth compared to other energy sources. Global energy consumption is projected to increase from 153 quads in 2022 to between 170 quads and 241 quads by 2050, presenting significant opportunities for unconventional gas exploration and development. In this chapter, we will focus on the unconventional gas resources that are sourced from the subsurface (i.e., CBM, tight gas, shale gas, methane gas hydrates, and hydrogen).

2. History of exploration and production of unconventional gas resources

The recognition of CBM is the earliest among unconventional gas resources due to the early utilization of coal during human civilization. However, commercial CBM production in North America starts in the 1970s [17]. CBM production or even exploration arises later in the beginning of the twenty-first century in developing countries following coal mining [9]. Exploration of tight gas and shale gas also rapidly developed in the 1990s, owing to successfully established technology and equipment for horizontal wells and hydraulic fracturing [18]. Production of tight gas provided a remarkable natural gas supply followed by CBM and shale gas before the 1990s. At the beginning of the twenty-first century, production of tight gas and CBM remained stable after slight increases around 1995 and 2005 [19]. After 2012, production of shale gas sped up and continuously increased, beyond that of tight gas and CBM [19]. This successful exploration and production of shale gas (also called the shale gas revolution) resulted in the United States becoming a net natural gas exporter in 2017. Exploration and production of shale gas attracted developing countries' attention since the shale gas revolution [20–22]. Exploration of gas hydrates develops slowly and currently remains in the stage of geological assessment [23]. Production of gas hydrate also stays in the stage of testing, urgently in need of breakthrough of equipment and production strategies [24–26]. The study of hydrogen recently started to attract increasing attention because of the carbon neutrality universal agreement at the Paris Climate Conference in 2015 [13]. Currently, commercial exploration and production of hydrogen have not started yet due to the immaturity of the hydrogen research, but these activities will arise along with a deeper understanding of hydrogen generation and accumulation in the following decades.

3. Theoretical and experimental insights on unconventional gas resources

3.1 Coalbed methane

CBM refers to mainly adsorbed methane that was generated and accumulated in coalbeds. Theoretical and experimental development in CBM mainly includes generation, accumulation, petrophysical characterization, and assessment.

Based on hydrocarbon's theories of organic origin, experimental simulation of organic matter (OM) thermal maturation was often employed to investigate the products of various types of OM (i.e., kerogen) at increasing thermal maturities. Type III OM generates CBM, including mainly methane (often >95%), minor C_{2+} hydrocarbons, carbon dioxide, and nitrogen [27]. Organic macerals mainly consist of vitrinite, inertia, and liptite. The composition of generated CBM is consistent with predominant methane regardless of the thermal maturity of OM [9]. Like conventional OM maturation, CBM can be either primary or secondary (due to tectonic uplifting) biogenetic when vitrinite reflectance (R_O) is lower than 0.6% or located in shallow depths where bacteria can reach. Also, CBM can be thermogenic when OM experiences thermal degradation or cracking when vitrinite reflectance (RO) is higher than 0.6%. Key compositional and isotopic parameters (e.g., C_1/C_{2+}, $\delta 13C_1$, δD) were often used in cross-plot plates to determine the CBM genesis [9, 28–30].

Different from conventional trap theories, self-generation and accumulation theories were proposed with no clear trap boundaries [31]. CBM mainly accumulates in stable structural low, for example, the center of a syncline. The richness of CBM decreases from the deep center to the shallow flanks and increases along with the coal rank (i.e., thermal maturity). According to high-pressure gas adsorption experiments, the storage states include mainly adsorbed state (accounting for 70–95%), free state (10–20%), and dissolved state (rare). CBM corresponds to the occurrence of coalbeds in Carboniferous-Permian (~40%), Triassic-Jurassic (~10%), and Cretaceous-Paleogene (~50%) due to the prosperity of terrestrial vascular plants after the Devonian [9].

Petrophysical characterization facilitates the theories of pore evolution. Experimental research mainly includes characterizations of coalbeds' matrix pore systems for gas storage and fracture networks for gas flow. Techniques in petrophysical characterization include low-pressure gas (e.g., N_2 and CO_2) adsorption isotherms, high-pressure mercury injection porosimetry (MIP), nano- or micro-computed tomography (CT), and field emission or focused ion beam scanning electron microscopy (FE/FIB-SEM). For CBM, the classification of pores proposed by Ходот [32] is often used based on industrial adsorbent: micropores (<10 nm), tiny pores (10–100 nm), mediate pores (100–1000 nm), and large pores (>1000 nm) [32]. Another classification of pores is also often adopted and proposed by the International Union of Pure and Applied Chemistry (IUPAC) based on mutual interaction between the adsorbent and adsorbate molecules in 1972: micropores (<2 nm), mesopores (2–50 nm), and macropores (>50 nm) [33]. The IUPAC classification is also widely used in shale. To keep consistency, we adopt the IUPAC classification in this chapter. At shallow depth ($R_o \leq 0.5\%$), primary macropores and mesopores mainly developed resulting in a high porosity. Along with the coal rank (e.g., bituminous coal, $R_o \sim 1.5\%$), primary micropores become dominant and possess the lowest pore volume and specific surface area [34]. With the coal rank increasing, secondary micropores and mesopores develop, resulting in a growth of microporosity. On the other hand, three major fracture types are widely developed in coalbeds: (1) endogenetic fractures (e.g., face and butt cleats) due to an internal tensile force created by compositional and structural variation, volume shrinkage, and coalification; (2) exogenetic fractures due to structural stress; and (3) inherited fractures showing both endogenetic and exogenetic features. The heterogeneity of fracture permeability is extremely high even in different blocks in the same basin. Fracture permeability decreases exponentially along with burial depth [9].

Efforts in precise assessment of CBM reserves started in the 1970s. A series of static and theoretical methods were employed to obtain the relationship between desorption and time based on in-situ measured desorbed data [35–39]. As such, the kinetic theories of gas desorption were being developed through canister desorption experiments (CDE) [3]. In shallow coalbeds, the gas-in-place (GIP) can be acquired based on the amount of desorbed gas, which is linearly proportional to the square root of the time of desorption based on CDE observation and data fitting. For deep CBM, a curve-fitting method by Yee et al. [39] was proposed based on Fick's diffusion theory [39]. Recently, a quasi-Langmuir kinetic theory was proposed for the relationship between desorption and emission time, providing a more precise and practical solution for estimating the CBM GIP [3].

3.2 Tight gas

Tight gas mainly refers to methane in tight sandstone formations, sometimes in tight carbonate formations, with a permeability of <0.1 mD and a porosity of <10% [40, 41]. Other alkane gases (C_{2-4}) and non-hydrocarbon gases (CO_2 and N_2) are in low volume fractions. Theoretical and experimental research on tight gas is based on previous conventional generation-migration-trapping theories of sandstone and carbonate.

Novel theoretical research on tight gas is the mechanisms of accumulation, for example, the time order of gas accumulation and reservoir tightness. As such, based on field observation and simulation experiments, Jiang et al. [42] divided tight gas reservoirs into two types: (1) reservoir tightness occurs before hydrocarbon peak generation, expulsion from source rocks, and charging into reservoirs; and (2) reservoir densification occurs after hydrocarbon charging and accumulation in reservoirs [42].

Recent studies also focused on the mobility of fluids and pore structure (i.e., storage space) in the theory of accumulation [43–45]. Low-field nuclear magnetic resonance (NMR) coupling high-pressure mercury injection porosimetry (MIP) is useful in determining the movable fluid capacity, heterogeneity of pore systems, and fractal characteristics of tight gas formations [44–45]. On the other hand, ultra-small-angle or small-angle neutron scattering (USANS/SANS) and low-pressure gas (N_2 and CO_2) adsorption isotherms for characterization of pore structure in shales were also employed in tight gas to reveal the pore size distribution of tight gas reservoirs [46, 47].

Based on reservoir characteristics and structural location, Dai et al. [7] observed two types of tight gas reservoirs and enriched tight gas' accumulation-trapping theory: (1) ones (called continuous-type) located in structural low (e.g., syncline flank) with reversed gas-water distribution (i.e., gas below water) and indistinct trap boundaries and (2) the others (called trap-type) located at the structural high of a trap with low porosity and permeability [7]. Usually, tight gas possesses a lower natural productivity than commercial gas production, requiring specific economic and technical conditions.

3.3 Shale gas

Shale gas refers to adsorbed, free, and dissolved natural gas (also mainly methane with minor C_{2-4} and non-hydrocarbon gases, CO_2, and N_2) in shale. Due to the tight and fine nature of shales, the heterogeneity of its composition and microstructure is extraordinarily high as it can contain the most common sedimentary minerals and develop abundant nanopores and micropores in OM and inorganic minerals [20, 48–49].

Theoretical and experimental research on shale gas dramatically increased at the beginning of the twenty-first century, mainly involving organic petrology, petrophysics, geomechanics, reserve assessment, and rock-fluid interactions during and after hydraulic fracturing.

As a major constituent of the generation theory, OM is critical for shale gas as a number of nanopores in OM provide extremely high specific surface area for adsorption sites. Since Curtis et al. [50] observed that the phenomenon of two adjunct organic particles developed pores differently under the same geological conditions, the effects of organic petrology on pore development in OM began to attract researchers' attention [50]. Loucks later suggested that OM (mainly kerogen) provided organic pores [8]. Bernard et al. [51] further proposed that solid pyrobitumen could also be another source of organic pores [51]. Recently, Gao et al. [52] separately observed different organic particles under oil-immerse reflect light microscopy and found that along with different thermal maturities, most organic particles would evolve to develop organic pores but be differently impacted by the same burial depth [50].

For the theory of pore development, sedimentary facies and diagenesis result in the petrophysical difference of the shale matrix due to spatial distribution and pore evolution of organic and inorganic particles [22, 53–55]. OM was long thought to be important as it will develop remarkable micropores and nanopores through thermal maturity [6, 56]. Different types of OM possess a different threshold of pore development, for example, for type II kerogen, the OM starts to generate pores when $R_o > 0.9\%$ [12]. Pore volume in OM will first increase and then decrease at the overmature stage due to compaction. Clay minerals were later regarded as another significant source of micropores and nanopores, especially in organic-lean shales [57, 58].

As for geomechanical theory, the compositional and spatial distribution of organic and inorganic particles determine the geomechanics of shales. For instance, Benge et al. [59] recognized clay-rich (clays, ~29 wt %) Caney Shale intervals in southern Oklahoma to be of a Poisson's ratio > 2.5, a low resistivity, and a high gamma [59]. The clay-rich ductile shale intervals experienced more creep behaviors than reservoir intervals (calcareous, with clays 13.5 wt%) under stress. Xiong et al. [60] followed research on Caney Shale's spatial distribution and found that the microscale quartz particles were embedded inside the flexible clay background for the clay-rich ductile shale intervals, explaining why this ductile interval would experience more creep behaviors than the rigid reservoir intervals in which most rigid particles contact each other with clays in the interparticle space [60].

Shale gas reserve assessment is commonly based on the estimation of GIP through experimental measurements of the gas-bearing capacity of cores. The occurrence of shale gas includes free gas that can be estimated by pore volume and state of equations, adsorbed gas that is estimated by specific surface area and adsorption prediction under reservoir conditions, and dissolved gas that is estimated from dissolution of methane in OM and fluids [61, 62]. In practice, dissolved gas was often thought of as negligible, and only the other two were calculated. The pore volume of cores can be obtained by helium porosity measurements or gas adsorption isotherms. Pore-specific surface area is usually measured by low-pressure nitrogen and carbon dioxide adsorption isotherms. In gas fields, another technical workflow was also employed to obtain the GIP. The fresh cores were once uplifted to the surface in situ and quickly placed in an air-tight canister that was filled up with an over-saturated NaCl solution for canister desorption testing, to obtain the desorbed gas [3]. The measured desorption-time relationship is used to predict the lost gas during the recovery of cores inside the wellbore. The cores after canister desorption testing are crushed to obtain the residual gas in the lab. The sum of desorbed gas, lost gas, and residual gas provides the GIP as well.

The theory of rock-fluid interactions along with shales' tight nature and creep behaviors may explain the fast decline of shale gas production curves. To obtain commercial production of shale gas, horizontal wells and hydraulic fracturing were often used in shale gas fields. However, over 50% of injected hydraulic fracturing fluids will remain in shale formations and react with shale matrices. Based on batch reactor analysis, Awejori et al. [11, 63] observed the shift of brittleness toward ductility on mineralogy and fracture walls for both short-term (interactions between hydraulic fracturing fluids and shale matrix) and long-term (interactions between mixed fracturing and formation fluids and shale matrix) [11, 63]. Pyrite is readily reacted with dissolved oxygen in the injected fracturing fluids, generating an acid setting that will dissolve the carbonates and feldspars and form fine shitty clay pieces.

3.4 Gas hydrates

Gas hydrates refer to ice-like clathrates, inclusive chemical compounds with ordered water molecules forming host cavities and gases (e.g., CH_4, CO_2, H_2S, and C_2H_6) as guest molecules. In nature, as part of generation theory, gas hydrates occur in permafrost regions and continental margins at a low-temperature and a high-pressure environment often with structure I (primary), structure II, and structure H. In a seismic profile, the gas hydrate-bearing formation can be often recognized by a bottom simulating reflector (BSR) that reveals the base of the hydrate. However, these BSR indications are not always accurate in determining the presence of hydrates [5]. Those structures represent different arrangements of water molecules, resulting in different cavities in size and geometry for guest molecules. Since gas hydrate was discovered in a lab in the early nineteenth century, gas hydrate has not yet been commercially exploited due to high operation costs, potential environmental risks, and the absence of production strategies and equipment. Recently, research on gas hydrates gave rise to potential production as an energy alternative because methane emission lowers the climate change issue, and causes submarine slope geohazards due to dissociation [5].

A unit volume of fully saturated gas hydrate under standard conditions provides ~164 m^3 methane, resulting in a high 50 probability reserve of over ~2000 trillion cubic meters [64]. Recently, production testing has been attempted in North America, Japan, Northern Europe, China, Korea, and India. However, the production of gas hydrates still did not last long due to the increasing production of sand and water along with recovery. Until now, the longest production testing can last for ~2 months [25].

If gas hydrates were melted by global warming and lease methane for a decade, the greenhouse gas amount would be 10 times that in the atmosphere [64]. Other scholars suggested that such release of hydrate-host methane may not be abrupt and massive as the found gas hydrates often locate a kilometer deep below the seafloor; however, it will be chronic and potentially contribute to a long-term greenhouse emission. Producing methane and storing CO_2 simultaneously in gas hydrates recently provided another alternative for satisfying both energy demand and carbon neutrality.

A few degrees Celsius warming up of a seafloor will result in a dissociation of shallow vulnerable gas hydrates. As the shallow gas hydrate often occurs in the pore space and serves as cement for loose sedimentary particles, the dissociation of gas hydrates will decrease the geomechanics of especially slope gas hydrates, causing instability (e.g., bulk change) and slope failure [23, 65].

3.5 Hydrogen

After a universal agreement on carbon neutrality at the Paris Climate Conference in 2015, hydrogen, together with carbon capture, utilization, and storage (CCUS) and geothermal resources, started to attract increasing attention of academia and industries in the 2020s [1, 4, 13]. Compared to methane, hydrogen was regarded as the better alternative after methane to economically utilize renewable resources for energy supply. A whole chain of hydrogen sectors is building up in the pioneering countries, involving generation, transportation, storage, and utilization of hydrogen at the research level. Overall, the exploration of hydrogen has just started at the assessment stage, and exploitation of hydrogen is also at the beginning stage except for some localized hydrogen usages (e.g., Bourakébougou in Mali, West Africa) [1]. Recently, theoretical and experimental insights on hydrogen mainly involve generation theories of hydrogen and subsurface geological storage of hydrogen.

In hydrogen generation theories, various techniques/processes are recognized to form all kinds of hydrogen nowadays, including electrolysis (green hydrogen), hydrocarbon reforming (gray/blue hydrogen), geological process (gold/white hydrogen), and anthropogenically geochemical stimulation (orange hydrogen). Natural hydrogen generated by geological processes (e.g., serpentinization) was thought to be promising with the lowest cost among all the techniques [1]. For instance, when water contacts ultramafic rocks (iron-rich and magnesium-rich olivine as the main constituent), olivine first experiences hydration, generating serpentine and brucite, which forms hydrogen through oxidation. Those interactions are common and closely related to plate boundaries, for example, transform faults, subduction zones, mid-ocean ridges, ophiolite complexes, and fold belts. Decomposition of OM, radiolysis, seepage from the deep mantle, and microbial metabolism can also produce hydrogen in geological processes.

Small quantities and frequent usage of hydrogen require storage by high-pressure tanks, metal hydrides, ammonia, and liquid organic hydrogen carriers. Due to high operation costs and low economic issues, the aforementioned techniques were not widely employed yet. Large-scale hydrogen storage uses subsurface salt caverns or depleted hydrocarbon reservoirs, with the advantages of low maintenance cost, established technology, and long-term integrity. To date, subsurface storage of hydrogen is still at a testing stage and only a couple of cases were successful, for instance, the subsurface hydrogen storage programs in depleted hydrocarbon reservoirs in Molasse Basin in Austria and Diadema region in Argentina.

4. Development of unconventional gas resources

Producing from CBM, shale gas, and tight gas reservoirs involves the hydraulic fracturing process conducted from horizontally drilled wells. Gas hydrate and hydrogen development and production, on the other hand, rely on different approaches. Thus, Section 4 is subdivided into three major sections: Section 4.1 will introduce the development of CBM, tight gas, and shale gas as a group because of similar technologies for their development strategies; we detailed their development surveillance methodology and simulation technologies in two subsections under Section 4.1. Following the above group, Section 4.2 and Section 4.3 will introduce the development of gas hydrate and the development of hydrogen, respectively.

4.1 Coalbed methane, tight gas, and shale gas

The challenges of CBM, tight gas, and shale gas development encompass environmental, political, and technical aspects. These include, but are not limited to, potential groundwater pollution, greenhouse gas emissions, market sensitivity to natural gas prices, high uncertainty in subsurface reserves, and pressure from carbon net-zero goals [66–68]. As of 2017, the United States, the largest CBM producer, generated 1.76 trillion cubic feet of methane, contributing 3.6% to its total dry gas production for the year [69]. Similar to shale gas development, CBM extraction heavily relies on advancements in horizontal drilling and hydraulic fracturing technologies [31, 70].

The integration of horizontal drilling and multi-stage hydraulic fracturing has unlocked the potential of unconventional gas reservoirs, especially low permeability reservoirs by creating intricate networks of highly permeable fractures within the rock formations [71, 72]. Proper drilling, well design, and completion parameters lead to effective well completion and cluster efficiency, optimizing gross-to-net fracture geometries and ultimately enhancing reservoir drainage and production. The key engineering factors are summarized in **Table 1**.

4.1.1 Coalbed methane, tight gas, and shale gas surveillance methodology

The critical questions in unconventional gas development include (1) ensuring effective hydraulic fracture geometry; (2) assessing the ability of current methods to generate propped fractures effectively; (3) determining if the created fractures enable multiple benches to drain simultaneously; (4) optimizing the fracturing sequence of multiple horizontal wells; and (5) evaluating the impact of proppant, fluid, and cluster designs on completion quality and efficiency [85]. Addressing these questions

Key completion parameters	Impacts	Typical values
Well spacing	Production drainage among neighbor wells [73].	400–600 ft. [74], 660 ft. [75]
Lateral length	Contact drainage between the well and the formation [76].	6700 ft. [75], 5000–12,500 ft. [77]
Stage spacing	Completion efficiency, number of clusters within a stage [78].	190–320 ft. [75], 150–300 ft. [79]
Average cluster number per stage	Completion efficiency, cluster efficiency [80].	6–10 [75]
Cluster spacing	Drainage between the fractures and cluster efficiency [80].	32 ft. [75], 10–20 ft. [79]
Pump rate	Facture net pressure, affects fracture geometries, and proppant carrying capacity [81]	100 bbl/ft. [75]
Proppant intensity	Propped fracture geometry [82]	1500–2500 lbs./ft. [79]
Fluid intensity	Gross fracture geometry and proppant carrying capacity [83]	35–80 bbl/ft. [79]
Proppant mesh	Proppant transport efficiency, effective fracture area, and fracture conductivity [84]	100 mesh [80]

Table 1.
Important completion parameters for tight gas reservoir development.

necessitates pilot field designs coupled with sophisticated surveillance methods. This section will introduce microseismic monitoring, distributed acoustic sensing, radioactive tracer, distributed temperature sensing, and time-lapse geochemistry in detail.

Microseismic monitoring records the shear slippage of rock bedding planes and natural fractures during hydraulic fracture propagation. The microseismic "events" recorded provide direct indications of the formed fracturing system. This application provides fracture half-length, height, asymmetry, azimuth, and volume estimates and is therefore commonly used for fracturing model calibration, as shown in the study by Cipolla and Li et al. [86, 87].

Distributed acoustic sensing (DAS) has been recently utilized for fracture hit and fracture geometry estimation. It can be implemented at treatment wells and nearby far-field vertical and horizontal wells. From near-wellbore measurement, during the fracturing process, the color in the LF-DAS plot represents the acoustic energy level (hotter represents a higher energy level), which indicates the cluster efficiency and fluid distributions. The far-field DAS takes low-frequency DAS (LF-DAS) data as the fiber strain rate changes during the fracturing. From horizontal treatment wells, the effectiveness of stimulation, zonal isolation, and cluster efficiency can be dynamically evaluated. If hydraulic fractures propagate and intersect with offset horizontal wells, a fracture hit occurs, providing insights into the half-length of the hydraulic fractures. If no intersection occurs, the fracture process zone can be evaluated instead [88]. The vertical fiber well is usually used to evaluate the fracture height growth and fracture tips (**Table 2**) [89].

The first implementation of down-hole DAS was conducted in February 2009 by Shell in Canada. The most important advantage of it over other classical surveillance tools is that DAS offers real-time and dynamic responses to advise the fracturing operation in place. An example of comprehensive DAS implementation practice can be found in the work of Manchanda et al. [80]. They implemented both near-wellbore horizontal treatment wells and far-field offset horizontal and vertical wells to evaluate the performance of stimulation work as a whole. The fracture heights upper limits, fracture gross geometries, and the fracture hits along the whole wellbore are diagnosed.

Radioactive tracers, utilizing various isotopes, are mixed into the fracturing fluid at different stages of the hydraulic fracturing process. Logging and imaging tools are then employed to assess the radioactivity of the near-wellbore area, providing insights into the wet fracture geometry and the distribution of proppants based on the varying radioactivity of the isotopes used. However, this indirect method may introduce significant uncertainties, as some compounds could settle during the process. These tracers can be recovered from nearby wells early in the production phase, indicating

DAS Installation	Information obtained
Horizontal treatment well (near-wellbore)	Effectiveness of stimulation, zonal isolation effectiveness, cluster efficiency, and dynamic change during the treatment [88].
Offset horizontal well (far-field)	Fracture hit, treatment well wetted fracture half-length board range [90], fracture process zone if minimal intersection [88].
Offset vertical well (far-field)	Treatment well fracture height [83], vertical growth of fracture tip [89]

Table 2.
DAS Installation location and associated obtained information summary.

inter-well communication. The extent of this communication elucidates the complexities of the fracture system and may also be influenced by the production sequence or strategy [85]. Radioactive tracer aligns well with novel concepts of development and operation, therefore, it plays an essential role in reservoir evaluation. It is also relatively cost-efficient and requires less effort for implementation and monitoring. However, the radioactive tracer may not be a reliable source of surveillance method at high temperature reservoirs.

Since the late 2000s, Distributed Temperature Sensing (DTS) deployment has been utilized to monitor the movement of fracturing fluid during and after hydraulic fracturing. This technology converts induced short pulses in the fiber into molecular vibrations. The resulting light signals are backscattered at various wavelengths. Temperature is evaluated based on the intensity ratio of high backscatter light wavelengths. DTS can effectively assess cluster efficiency, fracture height, and vertical reservoir coverage in hydraulic fracturing operations [90].

Time-lapse geochemistry (TLG) is a well-established technique for monitoring well drainage height, well interference, and completion effectiveness. It involves tracing and deriving information about reservoir fluids, completion fluids, and reservoir rock characteristics during production periods using geochemical analysis. This methodology is instrumental in allocating production across vertical sections. For example, geochemical interpretation can provide insights into the vertical coverage of wells that land in different zones, as well as the production contributions from various reservoir benches [85]. TLG is another relatively economical technique for drained rock volume (DRV) evaluation, and the established DRV variations caused by geological variations of the subsurface can be even captured, showing DRV's high resolution diagnostic feature. Thus, it has been widely used across the United States, especially in Powder River Basin Niobrara and Turner tight oil [92], Permian [85], and Permian Delaware Basin [91].

Other tools are not discussed in detail and are not limited to pressure gauges, production logging, wellbore imaging technology, and others. Most tools are applied in a combination manner with other surveillance technologies as they give multiple layers of confidence in the analysis [93]. Thus, the ensemble of surveillance tools sometimes requires high investments, rapid data processing, and gigantic visualization power, while the comprehensive interpretation of multi-dimensional measurement reduces the ambiguity of the measurement and interpretations from the single tool.

4.1.2 Coalbed methane, tight gas, and shale gas modeling and simulation

The development of coalbed methane, tight gas, and shale gas using simulations mainly evaluates the tight or low permeability formations. Thus, the modeling processes of integrated systems are usually convertible among these energy types.

To address critical development questions such as estimating original hydrocarbon volumes, understanding fracture drainage behaviors, optimizing completion practices, determining optimal landing depths, evaluating uncertainties, and deciphering production driving mechanisms, a modeling workflow is indispensable. This workflow is designed to represent the static reservoir subsurface and simulate its dynamic development processes. However, due to the complexity of the development chain and the diverse challenges posed by physics, there is a trade-off. This trade-off involves balancing the accurate representation of interconnected real-world physics problems with the computational capacity needed to construct an integrated workflow capable of analyzing these problems comprehensively and effectively.

The philosophy of integrated tools can be traced back to the original Integrated Asset Modeling (IAM), which covers the combined subsurface production phase and surface engineering modeling phase. This workflow attempt was pioneered by Amoco in the 1960s. Later, IAM was applied to a 3D black oil reservoir by Emanuel and Ranney [94]. The focus of this approach gradually transits from integrated modeling workflow including surface back to subsurface only right after the U.S. unconventional horizontal well hydraulic fracturing activity takes off. The hydraulic fracturing element is added to the picture.

The modern integrated geology and engineering simulation workflow has been recently adopted to overcome the barriers between geology, petrophysical aspects, and completion and production realms. Bringing geomodel, stress distribution model, fracturing simulation and reservoir simulation together enables a successful optimization process and reliable development strategy decision-making (**Figure 1**).

The initial step typically involves constructing a static geological model. This process integrates multiple well data from the geomodel with 3D seismic data to establish both structural and stratigraphic models. Subsequently, 1D well logs are used to populate the petrophysical and geomechanical properties onto the 3D geomodel, thereby creating reservoir grids. This comprehensive approach forms the foundation for subsequent dynamic reservoir simulations and analyses [95, 96]. The in-situ stress derived from 1D well log data typically utilizes sonic logs and density logs, which are calibrated using measurements such as Dipole Shear Sonic Imaging Tool (DIFT), core measurements, and other data sources. This process aims to accurately capture the stress distribution within the reservoir.

Following the creation of grids suitable for fracture propagation modeling and flow simulation through upscaling, the process typically involves a closed-loop iteration. This iterative approach refines geomechanical property calibration for fracture propagation models and reservoir property calibration for flow simulations.

These calibrations are informed and constrained by a wide array of surveillance tools utilized in previous stages. This iterative refinement ensures that the models

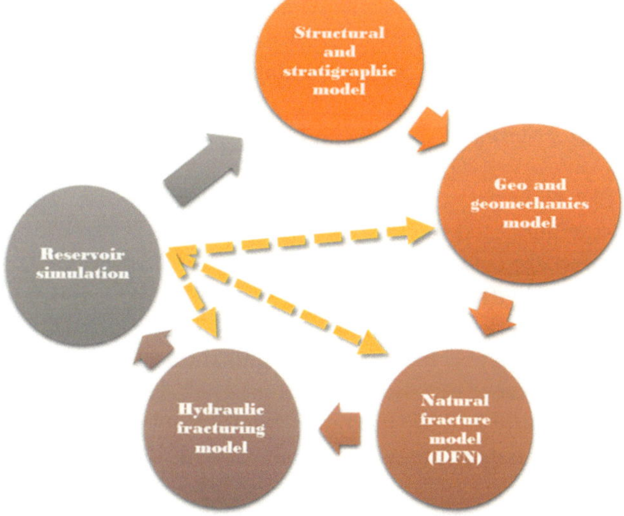

Figure 1.
Integrated workflow illustration.

accurately reflect the behavior of fractures and fluid flow within the reservoir, enhancing the reliability of predictions and optimization strategies.

To point out several examples of aforementioned applications, Wu et al. [97] used the G&G model-driven approach to evaluate the screen-out causes, fracture hit route cause, and search the solution to mitigate casing deformation from the 3D stress distribution and natural fracture characterization [97]. Mclntyre et al. [98] used a similar workflow to evaluate the effect of fracturing order to fracture height growth for stack development and further found the effect of completion parameters to return on investment (ROI) and net present value (NPV) of future child wells [98]. Li et al. [87] evaluated three wells' fracturing systems and thereafter calibrated the hydraulic fracture and activated natural fracture from microseismic signals, then the gas production performances and stimulated reservoir volumes of the three wells were carried out [87]. Han et al. [99] applied this workflow to capture the parent–child effects of wells on Hydraulic Fracture Test Site-2 [99]. The Estimated Ultimate Recovery (EUR) from various methods, such as calibrated simulation models, rate transient analysis, and decline curve analysis, is compared. Additionally, three completion designs were evaluated based on current well spacing using the calibrated model. It was concluded that the intermediate completion design effectively creates a robust fracture network within the well system.

The proposed workflow is then used to evaluate the pad development problems, including fracture-driven interaction, new well planning, and well planning. The representative works can be found in the studies by Wu and Xiong et al. [100–101]. There are abundant commercial and non-commercial software programs/codes capturing the discrete physics problem in **Figure 1**. However, there are only a few options to simulate the fracture propagation processes and reservoir production performance in a workflow manner. The commonly used platform involves the Resfrac, Zfrac-RE, and Kinetix-Intersect model by Schlumberger.

4.2 Gas hydrate

Due to the complex depositional environment of methane gas hydrates in Permafrost and deep-water areas, gas hydrate extraction is still challenging. The main approaches for natural gas hydrate recovery include depressurization, heating, inhibiting, CO_2 replacement, and solid exploitation.

4.2.1 Gas hydrate surveillance

Unlike the CBM, tight gas, and shale gas, the gas hydrate development is still not mature. The most widely used surveillance of gas hydrate exploration is seismic. It detects the structural characteristics of the reservoir and the gas hydrate-bearing zones. The presence of hydrate is usually indicated by the bottom simulating reflector. The bottom simulating reflector is caused by the high impedance mismatch between hydrate-bearing formation and free gas-bearing formation below the hydrate zone. The electromagnetic survey is another tool for gas hydrate exploration because the porous media natural fluid and the hydrate have different electromagnetic responses. The well logs and core analysis are usually used for hydrate formation analysis and characterization.

Determining the high-quality hydrate reservoir and evaluating the hydrate recovery rate rely on developing marine exploration methods. Even though the seismic survey for the exploration of hydrate has been adopted, the reliability of seismic results to the hydrate reservoir is still questionable. Developing the numerical simulation workflow for gas hydrate production is still urgent [102].

4.2.2 Gas hydrate simulations

The gas hydrate reservoir production involves fluid flow, heat transfer, and mass transport combined with physics processes. The solid phase vanishing could potentially happen during the process. Most of the previous simulation work mainly focuses on physics-behavior modeling. We summarized some of the pioneer physics-based numerical simulations of gas hydrate in **Table 3**.

Recently, the field-scale methane gas hydrate production simulation is arising. Taking the spatial variation of depositional layers and time-dependent thermal dynamic conditions into consideration, the UMSICHT HyReS is developed to simulate the methane production from hydrate reservoirs exchanging methane by CO_2. The commercial simulator STARS from CMG Ltd. can also simulate the aforementioned processes. To highlight a few representative recent production-level studies, Li et al. [107] numerically simulated the radial jet drilling technology for offshore South China Sea gas hydrate recovery [107]. Zhang et al. [108] evaluated gas production from gas hydrate deposits with multiple branch wells and evaluated the effect of the reservoir permeability, and initial hydrate saturation to gas production rate using the open-source code HydrateResSim (HRS) [108]. Myshakin et al. [109] used TOUGH + Hydrate code to build a 2D model to evaluate the gas production from the depressurization process in interbedded turbidite hydrate-bearing section in offshore India [109].

4.3 Hydrogen

Currently, 95% of hydrogen consumption originates from fossil fuels, especially from natural gas. Hydrogen production from natural gas requires a steam forming process, happening at extreme temperature conditions, i.e., 700–1000°C. Generating a unit mass weight of hydrogen emits 10 times the amount of CO_2 during this process [110]. In addition, the cost of generating hydrogen through steam reforming is three times higher than natural gas [111].

Hydrogen production simulation: The current hydrogen production simulation is mainly focusing on electrolysis hydrogen production. One example includes ProSimPlus software that can simulate the electrolysis of water and optimize the entire process. The simple approach is to build the material and energy balances by considering the possible recycling of electrolyte solutions and purges. Depending on

Researchers	Process	Main contributions
Sung et al. [103]	Depressurizing gas hydrate	3D multiphase finite difference numerical model
Kono et al. [104]	Depressurizing gas hydrate dissociation	Methane gas hydrate dissociation rate in porous medium
Jang and Santamarina [105]	Depressurizing and thermal stimulation	Coupled gas diffusion and fluid expansion behavior in a pore network modeling method
Xu et al. [106]	Hydrate decomposition	Methane hydrate decomposition in brine solution using molecular dynamics simulation

Table 3.
Examples of gas hydrate simulation at the physics fundamental level.

the operation condition and technologies, the efficiency and electrical features, such as voltage, current, and electrical power can be used. Other commercial software and non-commercial codes that model the hydrogen production simulation include the Aspen Plus model, H2A: Hydrogen analysis production model, H2FillS hydrogen filling simulator, and others.

5. Challenges in unconventional gas resources

Theoretically and experimentally, challenges in unconventional gas resources arise in different topics due to the level of exploitation. For CBM, the techniques, workflow, and criteria of exploration and production had been well established. The remaining challenges will be an accurate assessment of GIP and environmentally friendly enhancement of the recovery of CBM. Petroleum industries used to produce gas from sandstone. As such, research on tight gas mainly aims to study the gas mobility and enhance its recovery. Currently, the exploitation of shale gas is playing a major contribution to the natural gas supply in the United States. The remaining challenges of shale gas are to precisely assess its GIP and better understand microscale mechanisms of rock-fluid interactions, attempting to resist the fast production decline. As for gas hydrates, the major challenges are to precisely estimate its GIP and develop equipment and production strategies to stably and safely extract methane for a long period. The impacts of gas hydrates dissociation on marine and the atmosphere will also require continuous efforts. In the near future, economic generation, large-scale long-term storage, safe and sufficient transportation, and efficient usage of hydrogen will all be pressing challenges to be addressed by scientists and engineers together.

From a technical perspective, the reliable methods of gas hydrate reservoir identification, reservoir characterization, mechanical stability from hydrate dissociation, and natural gas transportation from hydrate recovery location remain unsolved. The challenge of hydrate recovery entails controlling the proper hydrate recovery rate, which leads to the stability of the economic potential. These recovery uncertainties are caused by the pore and stress stability issue due to the pore pressure change, permeability reduction, sediment softening and volume contraction, and segregated hydrate volume [63]. One of the challenges involved in hydrogen energy production is the high cost, low efficiency, and less durable electrolyzers; thus, advanced reforming and gasification are needed. The storage of hydrogen requires large-scale storage, including onsite bulk emergency supply and geologic formations. The economical dynamic response has yet to be realized to optimize storage strategies for co-locating stored hydrogen with end-use applications.

6. Opportunities in unconventional gas resources

Decarbonization for energy transition and carbon neutrality will be the trend for the following decades. Before the economic generation of hydrogen for fuel cells' usage, unconventional gas resources (mainly methane) will be the major gas supply to satisfy our annually increasing energy demand. Extraction of unconventional gas resources coupling with CCUS will be more preferred by energy and environmental sectors owing to low carbon footprint, encouraged bureautic policy, and motivated environmental responsibility. Replacing CBM and shale gas to enhance their recovery by injecting CO_2 had been tested in countries, such as the United States and China.

The same idea was also proposed for gas hydrates' recovery. CO_2 was also thought of as a cushion gas to keep the pressure of underground hydrogen storage stable.

Based on the U.S. Department of Energy's hydrogen program plan, the current near-term hydrogen production methods are mainly from gasification of coal, biomass, waste with carbon capture, utilization, and storage, advanced fossil and biomass reforming and conversion, and electrolysis. The long-term opportunities rely on advanced biological and microbial conversion and advanced thermo/photoelectrochemical H_2O splitting. From the delivery and storage perspective, the long-term opportunities are developing widespread pipeline transmission and distribution and chemical H_2 carriers, chemical H_2, and material-based storage.

7. Conclusions

A critical review, theoretical/experimental status, and challenges/opportunities of the primary unconventional gas resources, including coalbed methane, tight gas, shale gas, gas hydrate, and hydrogen were summarized and discussed in this chapter. Theoretical and experimental development of unconventional gas resources rapidly arose due to continuously increasing energy supply and will eventually evolve toward green and sustainable ones as well as their production. More challenges will be in deep and complex formations that previous exploration and production rarely reach either on land or at sea. Substantial opportunities will arise in the unconventional gas resources that will be obtained in sustainable ways and play important roles in the energy transition.

Acknowledgements

The project was supported by the "CUG Scholar" Scientific Research Funds at China University of Geosciences (Wuhan) (Project No. 2024019). The authors also thank the National Natural Science Foundation of China (Grant nos. 42130803 and 42072174) for their support. The authors acknowledge the support of the member companies of the University of Texas at Austin's Bureau of Economic Geology (BEG) Tight Oil Resources Assessment (TORA) industrial associates' program, as well as the support from BEG management.

Conflict of interest

The authors declare no conflict of interest.

Author details

Fengyang Xiong[1,2,3,4]*, Yiwen (Sophia) Gong[5] and Shu Jiang[1,2,3,4]

1 Hubei Provincial Key Laboratory of Oil and Gas Exploration and Development Theory and Technology, China University of Geosciences, Wuhan, Hubei, China

2 Key Laboratory of Tectonics and Petroleum Resources, China University of Geosciences, Wuhan, Hubei, China

3 School of Earth Resources, China University of Geosciences, Wuhan, Hubei, China

4 School of Sustainable Energy, China University of Geosciences, Wuhan, Hubei, China

5 Bureau of Economic Geology, Jackson School of Geosciences, The University of Texas at Austin, Austin, TX, USA

*Address all correspondence to: xfywilliam404@gmail.com

IntechOpen

© 2024 The Author(s). Licensee IntechOpen. This chapter is distributed under the terms of the Creative Commons Attribution License (http://creativecommons.org/licenses/by/4.0), which permits unrestricted use, distribution, and reproduction in any medium, provided the original work is properly cited.

References

[1] Blay-Roger R, Bach W, Bobadilla LF, Reina TR, Odriozola JA, Amils R, et al. Natural hydrogen in the energy transition: Fundamentals, promise, and enigmas. Renewable and Sustainable Energy Reviews. 2024;**189**:113888

[2] Santosh M, Groves DI, Yang CX. Habitable planet to sustainable civilization: Global climate change with related clean energy transition reliant on declining critical metal resources. Gondwana Research. 2024;**130**:220-233

[3] Xiong F, Hwang B, Jiang Z, James D, Lu H, Moortgat J. Kinetic emission of shale gas in saline water: Insights from experimental observation of gas shale in canister desorption testing. Fuel. 2021;**300**:121006

[4] Egerer J, Farhang-Damghani N, Grimm V, Runge P. The industry transformation from fossil fuels to hydrogen will reorganize value chains: Big picture and case studies for Germany. Applied Energy. 2024;**358**:122485

[5] Sloan ED. Fundamental principles and applications of natural gas hydrates. Nature. 2003;**426**(6964):353-359

[6] Ross DJK, Marc BR. The importance of shale composition and pore structure upon gas storage potential of shale gas reservoirs. Marine and Petroleum Geology. 2009;**26**(6):916-927

[7] Dai J, Ni Y, Wu X. Tight gas in China and its significance in exploration and exploitation. Petroleum Exploration and Development. 2012;**39**(3):277-284

[8] Loucks RG, Reed RM, Ruppel SC, Hammes U. Spectrum of pore types and networks in mudrocks and a descriptive classification for matrix-related mudrock pores. American Association of Petroleum Geologists. 2012;**96**(6):1071-1098

[9] Zou C. CBM. In: Unconventional Petroleum Geology. 2nd ed. Amsterdam: Elsevier; 2017. pp. 323-344

[10] Gong Y, Mehana M, Xiong F, Xu F, El-Monier I. Towards better estimations of rock mechanical properties integrating machine learning techniques for application to hydraulic fracturing. In: Proceedings—SPE Annual Technical Conference and Exhibition. 30 September – 2. Calgary, Alberta, Canada; 2019

[11] Awejori GA, Doughty C, Xiong F, Paronish T, Spycher N, Radonjic M. Integrated experimental and modeling study of geochemical reactions of simple fracturing fluids with Caney shale. Energy & Fuels. 2022;**36**(17):10064-10081

[12] Xiong F, Rother G, Radonjic M. Insights into controls of mineralogy and pore structure on the density of methane adsorption phase in shales under supercritical conditions. Energy & Fuels. 2022;**36**(17):10110-10122

[13] Hassan Q, Algburi S, Jaszczur M, Al-Jiboory AK, Al Musawi TJ, Ali BM, et al. Hydrogen role in energy transition: A comparative review. Process Safety and Environmental Protection. 2024;**184**:1069-1093

[14] Zheng L, Wei P, Zhang Z, Nie S, Lou X, Cui K, et al. Joint exploration and development: A self-salvation road to sustainable development of unconventional oil and gas resources. Natural Gas Industry B. 2017;**4**(6):477-490

[15] Speight JG. Unconventional gas. In: Natural Gas. 2nd ed. Houston: Gulf Professional Publishing; 2019. pp. 59-98

[16] US Energy Information Administration. International Energy Outlook. 2023

[17] Pan Z, Wood DA. Coalbed methane (CBM) exploration, reservoir characterization, production, and modeling: A collection of published research (2009-2015). Journal of Natural Gas Science and Engineering. 2015;**26**(Virtual special issue):1491-1494

[18] Middleton RS, Gupta R, Hyman JD, Viswanathan HS. The shale gas revolution: Barriers, sustainability, and emerging opportunities. Applied Energy. 2017;**199**:88-95

[19] US Energy Information Administration. Annual Energy Outlook. 2013

[20] Li Y, Jiang Z, Liu W, Huang R, Zhang K, Cheng S, et al. The effect of the tectonic thermal evolution history to gas content of lower Cambrian shales at Xiuwu Basin. Science Technology and Engineering. 2016;**16**(34):44-51

[21] Bi H, Jiang Z, Li J, Xiong F, Li P, Chen L. Ono-Kondo model for supercritical shale gas storage: A case study of Silurian Longmaxi shale in Southeast Chongqing, China. Energy & Fuels. 2017;**31**(3):2755-2764

[22] Xiong F, Jiang Z, Tang X, Li Z, Bi H, Li W, et al. Characteristics and origin of the heterogeneity of the lower Silurian Longmaxi marine shale in southeastern Chongqing, SW China. Journal of Natural Gas Science and Engineering. 2015;**27**:1389-1399

[23] Li Z, Deng Y, Rao S, Lu H. Methane gas bubbles affecting the formation and distribution of hydrates in kaolinite slit pores: A molecular dynamics study. Energy & Fuels. 2023;**37**(7):5102-5113

[24] Yamamoto K, Wang XX, Tamaki M, Suzuki K. The second offshore production of methane hydrate in the Nankai trough and gas production behavior from a heterogeneous methane hydrate reservoir. RSC Advance. 2019;**9**(45):25987-26013

[25] Li J, Ye J, Qin X, Qiu H, Wu N, Lu H, et al. The first offshore natural gas hydrate production test in South China Sea. China Geology. 2018;**1**(1):5-16

[26] Ye J, Qin X, Xie W, Lu H, Ma B, Qiu H, et al. The second natural gas hydrate production test in the South China Sea. China Geology. 2020;**3**(2):197-209

[27] Clayton JL. Geochemistry of coalbed gas—A review. International Journal of Coal Geology. 1998;**35**(1-4):159-173

[28] Bernard BB, Brooks JM, Sackett WM. Light hydrocarbons in recent Texas continental shelf and slope sediments. Journal of Geophysical Research Oceans. 1978;**83**(C8):4053-4061

[29] Schoell M. Genetic characterization of natural gases. American Association of Petroleum Geologists. 1983;**67**(12):2225-2238

[30] Whiticar MJ. Carbon and hydrogen isotope systematics of bacterial formation and oxidation of methane. Chemical Geology. 1999;**161**(1-3):291-314

[31] Zou C, Zhang G, Yang Z, Tao S, Hou L, Zhu R, et al. Concepts, characteristics, potential and technology of unconventional hydrocarbons: On unconventional petroleum geology. Petroleum Exploration and Development. 2013;**40**(4):413-428

[32] Ходот BB. Coal and Gas Outburst. Translated by Song S, Wang Y. Beijing: China Industry Press; 1966

[33] Everett DH. Manual of symbols and terminology for physicochemical quantities and units. Pure and Applied Chemistry. 1972;**31**(4):577-638

[34] Bustin RM, Clarkson CR. Geological controls on coalbed methane reservoir capacity and gas content. International Journal of Coal Geology. 1998;**38**(1-2):3-26

[35] Bertard C, Bruyet B, Gunther J. Determination of desorbable gas concentration of coal (direct method). International Journal of Rock Mechanics and Mining Sciences & Geomechanics Abstracts. 1970;**7**(1):43-65

[36] Kissell FN, McCulloch CM, Elder CH. The Direct Method of Determining Methane Content of Coalbeds for Ventilation Design. US Department of Interior: Bureau of Mines; 1973. p. 7767

[37] Kim AG. Estimating Methane Content of Bituminous Coalbeds from Adsorption Data. Department of the Interior: Bureau of Mines; 1977. p. 8245

[38] Smith DM, Williams FL. New technique for determining the methane content of coal. Proceeding of the Intersociety Energy Conversion Engineering Conference; (United States). 1981:2

[39] Yee D, Seidle JP, Hanson WB. Gas sorption on coal and measurement of gas content. In: Hydrocarbons from Coal. McLean: GeoScienceWorld; 1993

[40] Elkins LE. The Technology and Economics of Gas Recovery from Tight Sands. PTS: Society of Petroleum Engineers - SPE Production Technology Symposium; 1978

[41] Holditch SA. Tight gas sands. Journal of Petroleum Technology. 2006;**58**(06):86-93

[42] Jiang Z, Lin S, Pang X. The comparison of two types of tight sand gas reservoir. Petroleum Geology & Experiment. 2006;**28**(3):210-214

[43] Huang H, Chen L, Sun W, Xiong F, Ji W, Jia J, et al. Pore-throat structure and fractal characteristics of Shihezi formation tight gas sandstone in the Ordos Basin, China. Fractals. 2018;**26**(2):1840005

[44] Huang H, Li R, Xiong F, Hu H, Sun W, Jiang Z, et al. A method to probe the pore-throat structure of tight reservoirs based on low-field NMR: Insights from a cylindrical pore model. Marine and Petroleum Geology. 2020;**117**:104344

[45] Wang Z, Pan M, Shi Y, Liu L, Xiong F, Qin Z. Fractal analysis of Donghetang sandstones using NMR measurements. Energy & Fuels. 2018;**32**(3):2973-2982

[46] Huang H, Li R, Chen W, Chen L, Jiang Z, Xiong F, et al. Revisiting movable fluid space in tight fine-grained reservoirs: A case study from Shahejie shale in the Bohai Bay basin, NE China. Journal of Petroleum Science and Engineering. 2021;**207**:109170

[47] Clarkson CR, Freeman M, He L, Agamalian M, Melnichenko YB, Mastalerz M, et al. Characterization of tight gas reservoir pore structure using USANS/SANS and gas adsorption analysis. Fuel. 2012;**95**:371-385

[48] Xiong F, Jiang Z, Huang H, Wen M, Moortgat J. Mineralogy and gas content of upper Paleozoic Shanxi and Benxi shale formations in the Ordos Basin. Energy and Fuels. 2019;**33**(2):1061-1068

[49] Li P, Jiang Z, Zheng M, Bi H, Yuan Y, Xiong F. Prediction model for gas adsorption capacity of the lower Ganchaigou formation in the Qaidam Basin, China. Journal of Natural Gas Science and Engineering. 2016;**31**:493-502

[50] Curtis ME, Cardott BJ, Sondergeld CH, Rai CS. Development of organic porosity in the woodford shale with increasing thermal maturity. International Journal of Coal Geology. 2012;**103**:26-31

[51] Bernard S, Wirth R, Schreiber A, Schulz HM, Horsfield B. Formation of nanoporous pyrobitumen residues during maturation of the barnett shale (Fort Worth Basin). International Journal of Coal Geology. 2012;**103**:3-11

[52] Gao F, Wang C, Song Y, Hu Q, Wang C, Xiong F, et al. Organic pore heterogeneity and its formation mechanisms: Insights from the lower cretaceous lacustrine Shahezi shale in the Songliao Basin, NE China. Energy Exploration and Exploitation. 2022;**40**(1):112-131

[53] Tang X, Jiang Z, Huang H, Jiang S, Yang L, Xiong F, et al. Lithofacies characteristics and its effect on gas storage of the Silurian Longmaxi marine shale in the Southeast Sichuan Basin, China. Journal of Natural Gas Science and Engineering. 2016;**28**:338-346

[54] Gao F, Wang C, Song Y, Wan C, Xiong F, Li Z, et al. Quantitative characterization of organic pore structure from gas adsorption in lower cretaceous lacustrine shales in the Songliao Basin, NE China. Lithosphere. 2021;**2021**(Special 1):6644430

[55] Gao F, Song Y, Li Z, Xiong F, Chen L, Zhang Y, et al. Lithofacies and reservoir characteristics of the lower cretaceous continental Shahezi shale in the Changling fault depression of Songliao Basin, NE China. Marine and Petroleum Geology. 2018;**98**:401-421

[56] Zhang T, Ellis GS, Ruppel SC, Milliken K, Yang R. Effect of organic-matter type and thermal maturity on methane adsorption in shale-gas systems. Organic Geochemistry. 2012;**47**:120-131

[57] Clarkson CR, Solano N, Bustin RM, Bustin AMM, Chalmers GRL, He L, et al. Pore structure characterization of north American shale gas reservoirs using USANS/SANS, gas adsorption, and mercury intrusion. Fuel. 2013;**103**:606-616

[58] Xiong F, Jiang Z, Li P, Wang X, Bi H, Li Y, et al. Pore structure of transitional shales in the Ordos Basin, NW China: Effects of composition on gas storage capacity. Fuel. 2017;**206**:504-515

[59] Benge M, Lu Y, Katende A, Rutqvist J, Crandall D, Haecker A, et al. Connecting geomechanical properties with potential for proppant embedment and production decline for the emerging Caney shale, Oklahoma. In: SPE/AAPG/SEG Unconventional Resources Technology Conference, URTC 26-28 July. Houston, Texas, USA; 2021

[60] Xiong F, Wang Y, Puckette J, Grammer M, Awejori G, Radonjic M. Geochemical and microstructural characterization of shale rock: A workflow for site evaluation for energy-fluids storage. Energy & Fuels. 2024; Accepted

[61] Xiong F, Jiang Z, Chen J, Wang X, Huang Z, Liu G, et al. The role of the residual bitumen in the gas storage capacity of mature lacustrine shale: A case study of the Triassic Yanchang

shale, Ordos Basin, China. Marine and Petroleum Geology. 2016;**69**:205-215

[62] Xiong F, Rother G, Tomasko D, Pang W, Moortgat J. On the pressure and temperature dependence of adsorption densities and other thermodynamic properties in gas shales. Chemical Engineering Journal. 2020;**395**:124989

[63] Awejori GA, Xiong F, Katende A, Radonjic M. Produced fluid induced mineralogy and elemental alterations of Caney shale, southern Oklahoma. In: 57th US Rock Mechanics/Geomechanics Symposium. 25-28 June. Atlanta, Georgia, USA; 2023

[64] Bohannon J. Energy: Weighing the climate risks of an untapped fossil fuel. Science. 2008;**319**(5871):1753

[65] Lee JY, Ryu BJ, Yun TS, Lee J, Cho GC. Review on the gas hydrate development and production as a new energy resource. KSCE Journal of Civil Engineering. 2011;**15**(4):689-696

[66] Vengosh A, Jackson RB, Warner N, Darrah TH, Kondash A. A critical review of the risks to water resources from unconventional shale gas development and hydraulic fracturing in the United States. Environment Science Technology. 2014;**48**(15):8334-8348

[67] Meakin P, Huang H, Malthe-Sørenssen A, Thøgersen K. Shale gas: Opportunities and challenges. Environmental Geosciences. 2013;**20**(4):151-164

[68] Yap NT. Unconventional shale gas development: Challenges for environmental policy and EA practice. Impact Assessment and Project Appraisal. 2016;**34**(2):97-109

[69] Joshi D, Prajapati P, Sharma P, Sharma A. Past, present and future of coal bed methane (CBM): A review with special focus on the Indian scenario. International Journal of Coal Preparation and Utilization. 2023;**43**(2):377-402

[70] Tao S, Pan Z, Tang S, Chen S. Current status and geological conditions for the applicability of CBM drilling technologies in China: A review. International Journal of Coal Geology. 2019;**202**:95-108

[71] Birkle P. Recovery rates of fracturing fluids and provenance of produced water from hydraulic fracturing of Silurian Qusaiba hot shale, northern Saudi Arabia, with implications on fracture network. American Association of Petroleum Geologists. 2016;**100**(6):917-941

[72] Xiong F, Rother G, Gong Y, Moortgat J. Reexamining supercritical gas adsorption theories in nano-porous shales under geological conditions. Fuel. 2021;**287**:119454

[73] Liang B, Du M, Yanez PP. Subsurface well spacing optimization in the Permian Basin. Journal of Petroleum Science and Engineering. 2019;**174**:235-243

[74] Pei Y, Yu W, Sepehrnoori K, Gong Y, Xie H, Wu K. The influence of development target depletion on stress evolution and infill drilling of upside target in the Permian Basin. SPE Reservoir Evaluation & Engineering. 2021;**24**(03):570-589

[75] Zakhour N, Jones M, Zhao Y, Orsini K, Sahni V. HFTS-2 completions design and state-of-the-art diagnostics results. In: SPE/AAPG/SEG Unconventional Resources Technology Conference. 26-28 July. Houston, Texas, USA; 2021

[76] Alzahabi A, Alexandre Trindade A, Kamel A. Horizontal well completions using data analytics. Results in Engineering. 2023;**18**:101143

[77] Cui A, Cross T. The diminishing returns of extending lateral length across different basins. In: SPE/AAPG/SEG Unconventional Resources Technology Conference. 20-22 June. Houston, Texas, USA; 2022

[78] Han X, Feng FP, Zhang XC, Cao J, Zhang J, Suo Y, et al. An unequal fracturing stage spacing optimization model for hydraulic fracturing that considers cementing interface integrity. Petroleum Sciences. 2023;**20**(4):2165-2186

[79] Pei Y, Xiong H. Enhancing well performance and economic value via better well completion design—A Delaware Basin case study. SPE Journal. 2023;**28**(03):934-949

[80] Ibrahim AF. Optimizing cluster spacing in multistage hydraulically fractured shale gas wells: Balancing fracture interference and stress shadow impact. Journal of Petroleum Exploration and Production. 2024;**14**(7):2297-2313

[81] Zhao Y, Zhang Y, Wang C, Liu Q. Hydraulic fracturing characteristics and evaluation of fracturing effectiveness under different anisotropic angles and injection rates: An experimental investigation in absence of confining pressure. Journal of Natural Gas Science and Engineering. 2022;**97**:104343

[82] Gong Y, El-Monier I, Mehana M. Revealing the effect of bifurcated fractures and fluid parameters during multiphase slurry transport. Energy & Fuels. 2021;**35**(13):10519-10528

[83] Gong Y, Mehana M, El-Monier I, Viswanathan H. Proppant placement in complex fracture geometries: A computational fluid dynamics study. Journal of Natural Gas Science and Engineering. 2020;**79**:103295

[84] Tang Y, Ranjith PG, Perera MSA. Major factors influencing proppant behaviour and proppant-associated damage mechanisms during hydraulic fracturing. Acta Geotechnica. 2018;**13**(4):757-780

[85] Manchanda R, Liang Y, Meier H, Srinivasan K, Leonardi S, Johns M, et al. An integrated approach to development optimization using monitor wells and hydraulic fracture diagnostics in the Permian Basin. In: Society of Petroleum Engineers - SPE/AAPG/SEG Unconventional Resources Technology Conference. 13-15 June. Denver, Colorado, USA; 2023

[86] Cipolla CL, Wright CA. Diagnostic techniques to understand hydraulic fracturing: What? why? and how? SPE Production & Facilities. 2002;**17**(01):23-35

[87] Li Y, Zhang Q, Gong Y, Liu C, Leines-Artieda JA, Yu W. Calibration of complex discrete fracture network using microseismic events and fracture propagation modeling with seamless reservoir production simulation. Lithosphere. 2022;**2022**(Special 12):1409534

[88] Webster P, Cox B, Molenaar M. Developments in diagnostic tools for hydraulic fracture geometry analysis. In: Unconventional Resources Technology Conference 2013. 12-14 August. Denver, Colorado, USA; 2013

[89] Ciezobka J, Maity D. HFTS-2 (Final Report). Oak Ridge: U.S. Department of Energy Office of Scientific and Technical Information; 2022

[90] Shoaibi S, Farsi S, Suchta B, Persac S, Wakwak A, Gorgi S. Fluid and proppant distribution and fracture geometry calibration using permanent fiber-optic sensing DAS/DTS – Oman case study.

In: Society of Petroleum Engineers - SPE/IATMI Asia Pacific Oil and Gas Conference and Exhibition, APOG 2023. 10-12 October. Jakarta, Indonesia; 2023

[91] Jones P, Dressler D, Conner T, O'Brien J, Klaassen T, Bingham S. A novel approach to understanding multi-horizon fluid flow in unconventional Wells using produced water time-lapse geochemistry: Powder River Basin. In: Wyoming (American Association of Petroleum Geologists 9th Unconventional Resources Technology Conference); 26-28 July. Houston, Houston, USA; 2021

[92] Wang Y, Esson F, Michael E, et al. Applications of Time-Lapse Geochemistry (TLG) to Delaware Basin Field Development (American Association of Petroleum Geologists 9th Unconventional Resources Technology Conference); 26-28 July 2021; Houston, Texas, USA

[93] King GE, Rainbolt MF, Swanson C. Frac hit induced production losses: Evaluating root causes, damage location, possible prevention methods and success of remedial treatments. In: Proceedings—SPE Annual Technical Conference and Exhibition. 9-11 October. San Antonio, Texas, USA; 2017

[94] Emanuel AS, Ranney JC. Studies of offshore reservoir with an interfaced reservoir/piping network simulator. Journal of Petroleum Technology. 1981;**33**(03):399-406

[95] Oloo MA, Xie C, Oloo MA, Xie C. Three-dimensional reservoir modeling using stochastic simulation, a case study of an east African oil field. International Journal of Geosciences. 2018;**9**(4):214-235

[96] Pourpak H, Newby W, Taubert S, Al Marzooqi H, Baig MZ, Lefebvre-Prudencio A, et al. Integration of geology, geomechanics, and completion data in modeling for future well and completion optimization: An unconventional Diyab case study. In: SPE/AAPG/SEG Unconventional Resources Technology Conference. 26-28 July. Houston, Texas, USA; 2021

[97] Wu J, Zeng B, Wang L, Huang HY, Pan Y, Song Y, et al. 3D integrated G&G model-driven mitigation workflow on screenout, frac hits and casing deformation in ultradeep shale hydraulic fracturing. In: Society of Petroleum Engineers—SPE Argentina Exploration and Production of Unconventional Resources Symposium. 20-22 March. Buenos Aires, Argentina; 2023

[98] McIntyre R, Gaines J, Gardner S, Mediani M, Hammerquist C, Li X, et al. Predictive geologic and geomechanical models for pad development in the Delaware Basin. In: SPE/AAPG/SEG Unconventional Resources Technology Conference. 20-22 July. 2020; Virtual

[99] Han J, Zhao Y, Li H, Tan J, Bessa F, Liu S, et al. A comprehensive simulation study of hydraulic fracturing test site 2 (HFTS-2): Part II—Development optimization in the Delaware Basin using an integrated modeling workflow. In: Society of Petroleum Engineers—SPE/AAPG/SEG Unconventional Resources Technology Conference. 13-15 June. Denver, Colorado, USA; 2023

[100] Wu R, Liu X, Xu L, Kebert B, Singh A, Fusilier LB, et al. Numerical investigation of fracture driven interaction mitigation strategies and field development optimization. In: Society of Petroleum Engineers—SPE/AAPG/SEG Unconventional Resources Technology Conference. 17-19 June. Houston, Texas, USA; 2024

[101] Xiong H, Sarkar P, Yoon S. Optimization of parent-child well

spacing and completion design: A Delaware Basin case study. In: Society of Petroleum Engineers—SPE/AAPG/SEG Unconventional Resources Technology Conference. 17-19 June. Houston, Texas, USA; 2024

[102] Moridis GJ, Collett TS. Strategies for gas production from hydrate accumulations under various geological and reservoir conditions. In: TOUGH Symposium. 12-14 May. Berkeley, California, USA; 2003

[103] Sung W-M, Huh D-G, Ryu B-J, Lee H-S. Development and application of gas hydrate reservoir simulator based on depressurizing mechanism. Korean Journal of Chemical Engineering. 2000;**17**(3):344-350

[104] Ho K, Narasimhan S, Song F, Smith DH. Synthesis of methane gas hydrate in porous sediments and its dissociation by depressurizing. Powder Technology. 2002;**122**(2-3):239-246

[105] Jang J, Santamarina JC. Recoverable gas from hydrate-bearing sediments: Pore network model simulation and macroscale analyses. Journal of Geophysical Research. 2011;**116**(B8):B08202

[106] Xu J, Gu T, Sun Z, Li X, Wang X. Molecular dynamics study on the dissociation of methane hydrate via inorganic salts. Molecular Physics. 2016;**114**(1):34-43

[107] Li J, Huang Z, Li G, Huang Z, Dai J, Cheng K. Field test of radial jet drilling technology in a surface formation. Journal of Petroleum Science and Engineering. 2022;**218**:110928

[108] Zhang P, Tian S, Zhang Y, Li G, Zhang W, Khan WA, et al. Numerical simulation of gas recovery from natural gas hydrate using multi-branch wells: A three-dimensional model. Energy. 2021;**220**:119549

[109] Myshakin EM, Seol Y, Lin JS, Uchida S, Collett TS, Boswell R. Numerical simulations of depressurization-induced gas production from an interbedded turbidite gas hydrate-bearing sedimentary section in the offshore India: Site NGHP-02-16 (area-B). Marine and Petroleum Geology. 2019;**108**:619-638

[110] Chouhan K, Sinha S, Kumar S, Kumar S. Simulation of steam reforming of biogas in an industrial reformer for hydrogen production. International Journal of Hydrogen Energy. 2021;**46**(53):26809-26824

[111] Sanchez N, Ruiz R, Hacker V, Cobo M. Impact of bioethanol impurities on steam reforming for hydrogen production: A review. International Journal of Hydrogen Energy. 2020;**45**(21):11923-11942

Chapter 5

Perspective Chapter: Development Experience and Current Status of Tight Gas Reservoirs in China

Yunfan Wen, Lili Liu, Shiwei Wu and Yifeng Yan

Abstract

The development of tight gas reservoirs in China has made significant progress in recent years, becoming an important component of unconventional natural gas resources in the country. Tight gas resources are widely distributed in the Sichuan Basin, Ordos Basin, Songliao Basin, and other regions, with enormous resource potential. During the development process, China actively applied advanced horizontal well technology, multi-stage fracturing, and fine modeling techniques, significantly improving recovery rates and economic benefits. China's development experience shows that tailored technological applications and continuous innovation are key to overcoming complex geological challenges. In addition, China actively seeks international cooperation to accelerate industry development through technology introduction and experience exchange. However, the development of tight gas reservoirs still faces challenges such as high investment, difficulty in development, and environmental protection. In the future, China will continue to strengthen technological research and development, and emphasize green development and risk management, in order to achieve efficient and sustainable utilization of tight gas resources.

Keywords: Ordos Basin, tight gas, technological innovation, energy structure, sustainable development

1. Introduction

Tight gas in China is primarily found in regions such as the Tarim Basin, Ordos Basin, and Sichuan Basin. These areas feature complex geological structures characterized by dense reservoirs with low porosity and permeability, making traditional gas extraction techniques ineffective for tight gas. In recent years, due to the growing energy demand and the plateauing of conventional natural gas production, tight gas has become an essential component of China's unconventional natural gas sector. National policy support and technological advancements, such as horizontal drilling and hydraulic fracturing, have driven the commercialization of tight gas development [1, 2].

The development of tight gas holds significant strategic importance for China. In terms of energy security, it effectively expands the supply of natural gas resources, reduces dependence on imported natural gas, and strengthens national energy

IntechOpen

security. From an environmental perspective, natural gas is considered a relatively clean fossil fuel, contributing positively to reducing carbon emissions. The development of tight gas aids in achieving the nation's energy conservation and emission reduction goals. Economically, by developing new energy resources, it provides employment opportunities, stimulates related technologies and industries, and contributes to regional economic growth.

In terms of economic evaluation, tight gas development faces challenges such as high initial investment and technical difficulty. However, as technology matures, large-scale production is realized, and market demand for natural gas increases, the economic viability of tight gas development is gradually improving. Policy support, such as tax incentives and subsidies, further enhances its economic feasibility. Additionally, development costs are continuously decreasing, thereby improving the profitability of tight gas projects [3, 4].

Comparing China's tight gas development with that of other countries reveals significant differences in development models, economic viability, and strategic approaches. China's tight gas development primarily relies on state-led energy companies such as China National Petroleum Corporation (CNPC) and Sinopec. Government policy support is a major driving force. Moreover, the ongoing advancement of localized technologies has helped overcome many engineering challenges. Economic viability is influenced by domestic policies, technological levels, and market demand. Initial investments and technological costs are high, and profitability depends on large-scale development. The strategy for tight gas development focuses on energy security, environmental protection, and economic growth. Policies are clearly directed, with an emphasis on long-term benefits.

In contrast, other countries like the United States primarily depend on market-driven approaches for tight gas development. The success factors include private enterprises, technological innovation, and support from capital markets. In these aspects, the operational efficiency and technological standards in the U.S. are relatively higher. Countries like the United States possess mature market mechanisms and technological innovations, leading to lower development costs, shorter investment return cycles, and strong profitability. The U.S. development strategy not only focuses on the domestic market but also actively expands into international markets, exporting natural gas and related technical services. Additionally, the market sensitivity of private capital allows for a more flexible and dynamic strategy (**Table 1**).

Overall, China possesses immense potential and market demand for tight gas development but still needs to focus on technological innovation, cost control, and optimizing market mechanisms to better meet the strategic energy needs and economic development goals of the country. Therefore, a detailed examination of China's experiences and current status in tight gas reservoir development can provide an opportunity for reflection and insight, and also offer substantial support for the country's energy strategy.

2. Distribution of tight gas resources

2.1 Ordos Basin: Core area of tight gas resources

The Ordos Basin is one of China's key areas for tight gas resources and a globally renowned hub for these resources. Located in central-western China, the basin spans approximately 370,000 square kilometers, making it the second-largest sedimentary

Comparison Indicators	San Juan Basin	Montney Basin	Ordos Basin	Sichuan Basin
Sedimentary	Coastal Plain Sandbar	Eolian Sand of Coastal Plains	Braided River	Fluvial Deposits in Continental Environments
Layer Thickness(m)	40–100	60–80	3–5	10–32
Layer Distribution	Stable Distribution		Small and Scattered	Widely
Crack Development	Localized Development		Underdeveloped	
Porosity(%)	3–12	3–8	3–14	9–16
Permeability(mD)	0.001–0.1	0.001–0.03	0.001–0.1	0.1–16
Burial depth(m)	750–2650	2100–3000	2800–3700	1300–3000
Gas Saturation	>60%	>70%	55–60%	30–60%
Resource Abundance($10^8 m^3 \cdot km^{-2}$)	>5.00	6.00–9.00	1.20	2.31
Cumulative Gas Production per Well	Vertical(0.20–1.00) × 10^8 m^3 Horizontal>1.00 × 10^8 m^3		Vertical(0.20–0.30) × 10^8 m^3 Horizontal(0.60–1.26) × 10^8 m^3	

Table 1.
Comparison table of the development of tight gas in China and the United States.

basin in the country. Characterized by typical cratonic basin geological features, its reservoirs mainly consist of sedimentary rocks from the Mesozoic to Paleozoic eras. The tight gas within the basin is predominantly stored in sandstone and mudstone layers with low porosity and low permeability. The reservoirs in the Ordos Basin typically exhibit porosity of less than 10% and permeability below 0.1 millidarcy, leading to poor gas flow. The increase in production relies on complex enhancement technologies. The reservoirs are characterized by their dispersion and fragmentation, resulting in low single-well yields, necessitating effective development through dense well networks and hydraulic fracturing.

The development of tight gas in the Ordos Basin is primarily concentrated in fields such as the Jingbian Gas Field, Dongsheng Gas Field, and the Sulige Gas Field. Among these, the Sulige Gas Field is one of China's largest integrated natural gas fields, with significant proven tight gas reserves, playing a strategic role in China's natural gas supply (**Figure 1**).

Development of tight gas in the Ordos Basin faces multiple technical challenges, including well network layout, hydraulic fracturing, and large-scale reservoir transformation technologies [1]. In recent years, China has implemented numerous technological innovations, such as horizontal drilling, multistage fracturing, and fracture network fracturing technologies. These advancements provide critical support for the efficient development of tight gas. The tight gas resources in the Ordos Basin serve as an important supplement to China's natural gas supply, with long-term development potential. The development of tight gas has invigorated the local economy, boosting infrastructure development and related industries. Compared to other fossil fuels, natural gas emits less carbon, and the development of tight gas aids the country in achieving its energy-saving and emission-reduction goals (**Figure 2**).

The development of tight gas in the Ordos Basin is crucial not only for enhancing China's energy self-sufficiency but also for contributing positively to regional economic growth and environmental protection.

Figure 1.
Sulige gas field gathering station.

2.2 Sichuan, Songliao Basin: Resource replacement area

The Sichuan and Songliao Basins are two key resource areas in China, renowned for their abundant oil and gas resources. In the Sichuan Basin, tight gas is primarily located within Jurassic and Triassic formations, and these tight sandstone reservoirs are characterized by low porosity and low permeability. This makes it one of the richest regions for tight gas resources in China, with considerable development potential. Hydraulic fracturing and horizontal drilling technologies are widely used in the Sichuan Basin to enhance reservoir permeability and increase gas production.

In the Songliao Basin, tight gas resources are mainly found within Cretaceous sandstone formations. Compared to conventional gas reservoirs, these formations are denser and more complex [3, 5]. As a significant gas source region in Northeast China, the tight gas resources in the Songliao Basin hold substantial exploration and development potential. Advanced fracturing technology and improved well completion techniques are applied to enhance single-well output and recovery rates, addressing the unique characteristics of the tight gas reservoirs.

Both the Sichuan and Songliao Basins hold significant strategic importance in resource extraction and technological advancement, playing a crucial role in ensuring China's energy security and sustainable development. Both regions face challenges related to replacing old with new resources. As traditional oil and gas fields gradually enter the decline phase, identifying and developing new reserves has become critical. Therefore, these regions are continuously engaging in geological exploration and technological innovation to seek new energy developments and alternative resources.

Figure 2.
Drilling and hydraulic fracturing sites at sulige gas field.

2.3 Tarim and Junggar Basin: Breakthrough area

The Tarim Basin and Junggar Basin are significant natural gas resource areas in western China. In recent years, the development of tight gas has become a crucial direction for resource succession. In the Tarim Basin, tight gas is primarily found in Paleozoic carbonate rocks and Mesozoic clastic rocks. The geological conditions are complex, featuring multi-layered and multi-type gas reservoirs [3, 6]. The tight sandstone and carbonate reservoirs are characterized by low porosity and low permeability. The main tight gas resources are concentrated in regions such as Tazhong, Tabei, and Kuqa, which have been confirmed to possess abundant tight gas reserves. The development of tight gas is most active in the Tazhong area, utilizing advanced hydraulic fracturing and multistage fracturing technologies to address reservoir tightness and enhance single-well output. Multiple tight gas production wells have been established, with production capacity gradually increasing.

In the Junggar Basin, tight gas is mainly present in the Jurassic and Upper Permian tight sandstones, also featuring low porosity and low permeability. The geological conditions vary, and as exploration deepens, the resource potential of different areas is gradually becoming evident. Tight gas resources are primarily distributed in regions such as the Mahu Depression and Changji Depression. The application of horizontal drilling and fracturing technology in the Junggar Basin has strengthened the extraction of low-permeability reservoirs. In recent years, certain development progress has been achieved in the Mahu area, with an increasing number of projects.

The development of tight gas resources in both basins faces a situation of challenges and opportunities. Although the reservoir conditions are complex, significant progress has been made through technological advancements, such as horizontal drilling and hydraulic fracturing. Developing tight gas resources not only effectively supplements the shortfall in conventional natural gas supply but also provides essential support for regional economic development and energy security. In the future, the potential of tight gas in these areas awaits further exploration and technological development.

2.4 Turpan-Hami, Bohai Bay: Future prospects

The Tuha Basin and Bohai Bay Basin are key oil and gas resource areas in China, and the development of tight gas resources within these basins is gradually progressing [7]. In the Tuha Basin, tight gas resources are mainly found in the Triassic, Jurassic, and Cretaceous tight sandstone formations, characterized by their typical low porosity and low permeability. The tight gas in the Tuha Basin is primarily concentrated in several depressions along the southern and northern fringes, such as the Toksun Depression and Wulatai Depression. The Tuha Oilfield Company is actively engaged in exploring and developing tight gas resources, employing advanced fracturing technologies to enhance permeability. Despite the high development difficulty, technological innovations have enabled several gas wells to achieve commercial production.

The Bohai Bay Basin boasts abundant tight gas resources, predominantly within Quaternary and Tertiary sandstone reservoirs. The high-pressure and low-permeability reservoir characteristics suggest significant development potential. These tight gas resources are mainly distributed in the western depressions of the Bohai Bay Basin, including regions such as Huanghua Depression, Jiyang Depression, and Liaoxi Depression. The area has attracted numerous oil and gas companies, such as PetroChina and Sinopec, which have made significant progress using horizontal drilling and multistage fracturing technologies. Multiple high-yield tight gas fields have been discovered in the Bohai Bay region, steadily increasing production capacity.

The development of tight gas in the Tuha Basin and Bohai Bay Basin faces technical challenges, but the prospects for these resources remain promising. Continuous technological improvements, particularly breakthroughs in drilling and fracturing technologies, have shown initial success in the commercialization of tight gas resources in these areas. With further advancements, these tight gas resources are expected to contribute more significantly to China's energy supply and economic development.

3. Key technologies for tight gas development

3.1 Multistage fracturing

Multistage fracturing is a complex enhancement technology primarily used for the development of unconventional oil and gas resources, such as shale gas, tight oil, and tight gas. This technique aims to maximize the contact area between the reservoir and the wellbore, thereby increasing resource output. The goal of multistage fracturing is to simultaneously fracture multiple segments at different depths in horizontal or vertical wells to create multiple high-conductivity channels. This approach significantly increases the flow area of the reservoir, enhancing the production of oil and gas wells. Multistage fracturing can significantly boost the productivity of oil and gas wells because it more fully exploits the reservoir. This technology is applicable not only to horizontal wells but also to vertical wells with multiple productive layers, especially under unconventional reservoir conditions. Through precise geological and engineering design, fractures can be accurately located and optimized, improving overall extraction efficiency. However, multistage fracturing involves complex engineering techniques, requiring high-precision equipment and technical operations. Compared to traditional fracturing techniques, it is more costly and demands more technical input and professional support. It requires large amounts of water and chemicals, which must be carefully managed to minimize environmental impact. In North America's shale gas and tight oil fields, multistage fracturing technology has been widely applied, and its success has provided insights and impetus for the development of unconventional resources in other regions globally. In recent years, this technology has also been gradually applied and has achieved results in the development of tight gas and shale gas in China (**Table 2**).

In the Sichuan Basin, multistage fracturing technology is widely used in tight sandstone reservoirs to increase single-well production. By using segmsented fracturing, the number and extension area of fractures can be maximized, thereby improving the recovery rate of gas wells. In response to the complex stratigraphic conditions of the Ordos Basin, multistage fracturing combined with horizontal well technology allows multiple fractures to effectively communicate with the reservoir, significantly enhancing development efficiency [8–10]. In the Songliao Basin, multistage fracturing technology has been applied to the transformation of tight gas wells in many areas, where multistage fracturing in horizontal well sections has improved gas production stability and yield.

With the continuous advancement of technology and accumulation of development experience, the application prospects of multistage fracturing in China's tight gas development are becoming broader. Ongoing technological innovation and meticulous management will drive the further application of this technology in the development of tight gas and other unconventional resources, achieving more efficient and sustainable resource utilization.

3.2 Well pattern design optimization

Optimization of well patterns and types is a crucial process in oil and gas field development. This involves designing and adjusting the arrangement, spacing, and types of wells based on geological conditions, reservoir characteristics, and economic demands to maximize development efficiency and economic benefits. This optimization is particularly important in the development of unconventional oil and gas reservoirs such as tight gas and shale gas. The primary goals of well pattern and type optimization are to increase the coverage range of the well network and the effective length of the wells to extract as much oil and gas resources from the reservoir as possible. It aims to minimize unnecessary drilling and fracturing operations, optimize the cost-to-benefit ratio, and ensure economic viability while protecting

Technical series	Advantages	Tools
Soluble Bridge Plug Staged Fracturing	The number of fracturing stages is not limited, the wellbore has full diameter, and high-volume injection is performed.	
Continuous Oil Pipe Layered Fracturing	Full bore of the well, precise layer fracturing, assembly line fracturing.	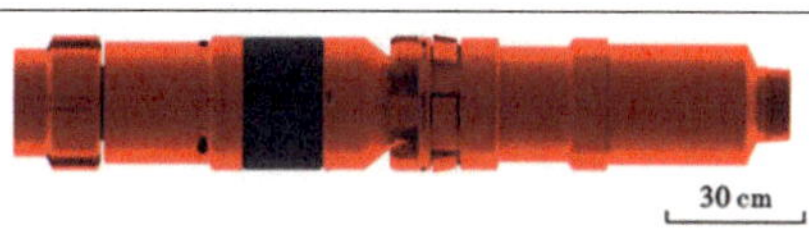

Table 2.
Multi-level fracturing stages.

the structural integrity of the formations and reducing environmental impacts for sustainable development. When optimizing well patterns and types, it is essential to consider geological characteristics such as reservoir thickness, porosity, and permeability. Changes in formation pressure and properties affect the selection and arrangement of well types. Reservoir heterogeneity and fracture distribution characteristics provide the basis for selecting well types and placements, while the production layer distribution and connectivity determine well spacing and layout. Common well types and layouts include horizontal wells, directional wells, and vertical wells. Numerical simulation, data analysis, and experimental testing results are commonly used in the optimization process [1, 11, 12]. The optimized well pattern and type design must be continuously evaluated and adjusted based on real-time production data during actual development. For unconventional oil and gas resources, as technology advances and geological understanding deepens, well pattern and type optimization is a dynamic process that requires ongoing refinement. The ultimate goal is to achieve efficient and economical resource development while protecting the ecological environment (**Figure 3**).

Well pattern and type optimization has played a significant role in the development of tight gas in China by designing and adjusting well layouts to increase resource recovery rates, reduce development costs, and maximize economic benefits. In the Sichuan Basin, dense well placements combined with horizontal wells and multistage fracturing technology have been employed to increase the contact area with complex sandstone reservoirs. In addition, geological modeling and numerical simulation have been used to optimize well locations and spacing, achieving a higher degree of reservoir utilization. In the Ordos Basin, a strategy of closely spaced, broadly covering horizontal well networks was applied to accommodate the characteristics of thin reservoirs and high heterogeneity in the region. In areas with lower reservoir pressure, dense well networks combined with continuous fracturing technology were employed to enhance gas flow capability. In the Songliao Basin, the multilayer well network layout was used to develop multiple production layers simultaneously for local tight sandstone, improving resource utilization efficiency. Advanced geophysical logging techniques and fracture monitoring tests were utilized to adjust the well network layout in real time, optimizing development plans. The successful experience of tight gas development in China provides a valuable reference for the development of other unconventional energy sources, with well pattern and type optimization being a key pillar for efficient and economic development.

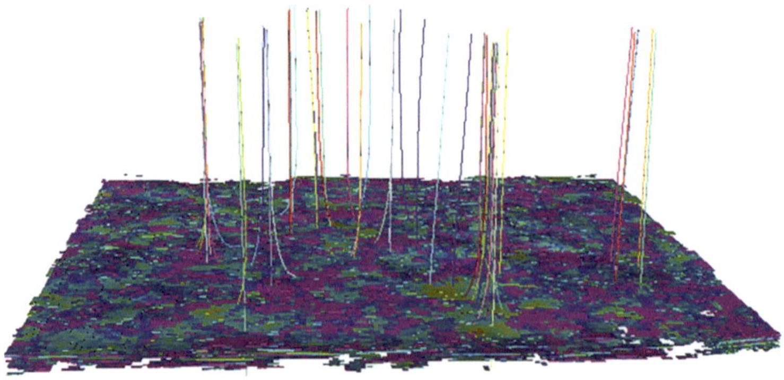

Figure 3.
Well deployment optimization simulation diagram.

3.3 Precision modeling

Fine-scale modeling technology is crucial in the development of tight gas resources in China. As it accurately describes complex geological features and reservoir characteristics, it plays a significant role in enhancing resource recovery rates, optimizing production plans, and reducing development risks [8, 9]. China's tight gas reservoirs are often situated in complex structural and depositional environments, exhibiting highly variable reservoir characteristics and intricate geological structures. These reservoirs are typically characterized by low porosity and permeability, making it challenging for traditional development methods to yield ideal production rates (**Figure 4**).

In the Ordos Basin, fine-scale reservoir modeling accurately depicted the thickness and distribution of mudstone interlayers and their impact on gas flow. This modeling provided key insights for horizontal well placement and fracturing design, thereby enhancing production capacity. In the Sichuan Basin, the application of fine-scale modeling analyzed the micro-pore structure and spatial distribution within tight gas sandstone reservoirs. Combined with three-dimensional seismic data, the model optimized well spacing and development strategies, significantly increasing single-well

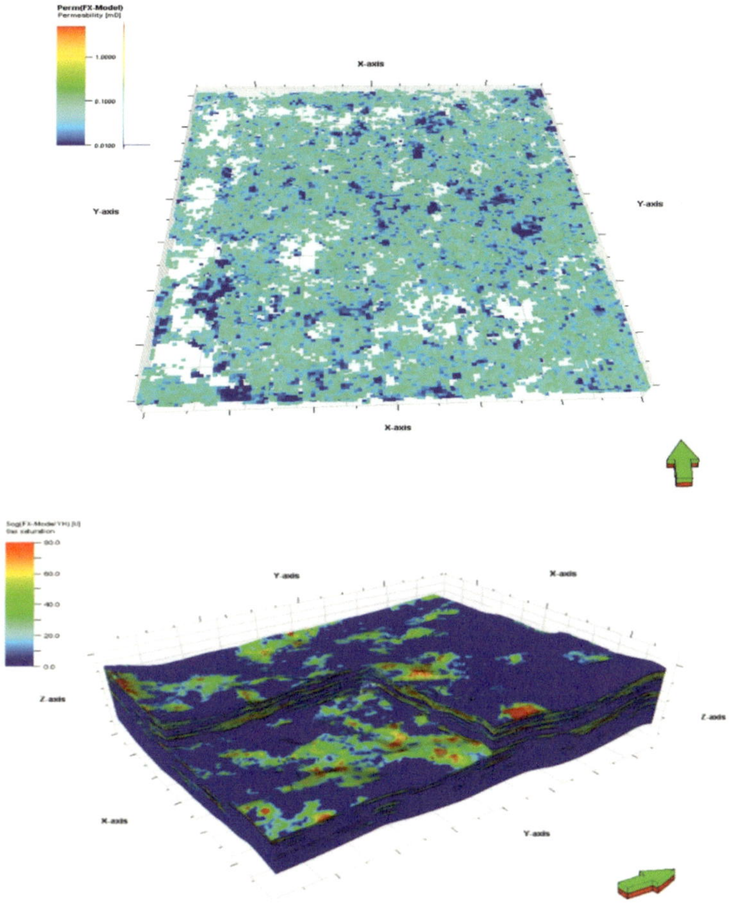

Figure 4.
Geological refinement modeling geological model.

output. In the Songliao Basin, by simulating different fracturing scenarios, the fracturing design was optimized to achieve higher extraction efficiency. Furthermore, in reservoir characterization, fine-scale modeling assisted in identifying complex fracture networks, improving fracture prediction accuracy.

Looking ahead, with continuous innovation and development, the application of fine-scale modeling will play a greater role in promoting the efficient and sustainable development of China's tight gas resources.

4. Development prospects

Tight gas is one of the important types of unconventional natural gas and is widely distributed in major oil and gas basins around the world, with a total resource of 210 trillion cubic meters. The Asia-Pacific and Americas regions have the richest resources, accounting for over 60% of the global total. Tight gas resources in China are relatively abundant, making up about 10% of the global total, but their distribution is highly uneven. The Ordos Basin is the main area for tight gas distribution in China.

In China, tight gas resources are widely distributed across several large basins, such as the Sichuan Basin, Ordos Basin, Tarim Basin, and Songliao Basin. These areas have diverse geological conditions, providing abundant tight gas reserves. Recent exploration activities indicate that China's tight gas reserves are estimated to reach trillions of cubic meters, offering significant development potential. The Chinese government actively promotes the development of unconventional natural gas resources and has implemented a series of policies to support the growth of tight gas, including financial subsidies, tax incentives, and support for technological innovation. Under the impetus of energy structure adjustment and carbon reduction goals, the importance of natural gas as a clean energy source is increasingly highlighted, with policy support favoring the development and application of natural gas, particularly tight gas. In recent years, advanced technologies such as horizontal drilling, multistage fracturing, and precise modeling have matured in the development of tight gas, significantly improving development efficiency and economic benefits. Technological advancements and large-scale development have steadily reduced extraction costs, paving the way for the economic development of tight gas. With economic growth and heightened environmental awareness, the demand for natural gas continues to rise, providing strong market momentum for tight gas development. China is committed to increasing the proportion of natural gas in its primary energy mix, and tight gas, as an important supplementary resource, plays a positive role in improving the supply structure.

China's tight gas holds significant potential for future production growth and development. Three key aspects will ensure the medium- to long-term development of tight gas in the country: utilizing untapped proven reserves, maximizing the potential of already tapped proven reserves, and effectively developing newly proven reserves. Maximizing already tapped proven reserves and developing untapped proven reserves can support a peak annual production of (700 to 800) $\times 10^8$ cubic meters, with the peak expected between 2030 and 2035 and stable production for more than 10 years. The development of newly proven reserves will further sustain stable production of tight gas through 2050 to 2060 (**Figure 5**).

China is also beginning to seek more international cooperation in the field of tight gas development, accelerating technological advancement and market expansion through collaborative research and experience sharing. Continuous breakthroughs in

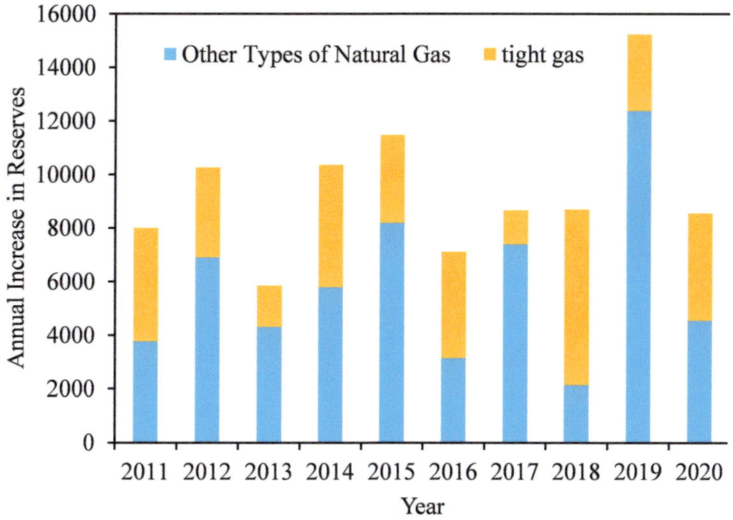

Figure 5.
Annual increase in proven reserves of natural gas and tight gas in China.

key technologies will be the core driving force for deepening tight gas development. The application of artificial intelligence and big data in oil and gas field development is becoming increasingly common. In addition to enhancing economic viability, environmentally friendly extraction technologies and green development concepts will promote the sustainable development of tight gas. Overall, the prospects for tight gas development in China are optimistic, but challenges in technology and market aspects remain. In the future, through policy support, technological innovation, and market drive, tight gas is expected to occupy a more important position in China's energy structure.

Author details

Yunfan Wen[1*], Lili Liu[1], Shiwei Wu[2] and Yifeng Yan[3]

1 PetroChina Exploration Development Institute of Changqing Oilfield Company, Xi'an City, Shaanxi Province, China

2 Engineering Technology Supervision Center of PetroChina Changqing Oilfield Company, Yan'an City, Shaanxi Province, China

3 Third Gas Production Plant of PetroChina Changqing Oilfield Company, Yan'an City, Shaanxi Province, China

*Address all correspondence to: 191216757@qq.com

IntechOpen

© 2025 The Author(s). Licensee IntechOpen. This chapter is distributed under the terms of the Creative Commons Attribution License (http://creativecommons.org/licenses/by/4.0), which permits unrestricted use, distribution, and reproduction in any medium, provided the original work is properly cited.

References

[1] Shiheng Y. Research on Productivity Prediction of Tight Gas Reservoirs Based on Machine Learning. Beijing: China University of Petroleum; 2023. DOI: 10.27643/d.cnki.gsybu.2023.002080

[2] Hanghang L. Dynamic Analysis and Productivity Evaluation of Tight Sandstone Gas Development. Xi'an Shiyou University; 2022. DOI: 10.27400/d.cnki.gxasc.2022.001036

[3] Ailin J, Yunsheng W, Zhi G, et al. Current status and prospects of tight sandstone gas development in China. Natural Gas Industry. 2022;**42**(01):83-92

[4] Zhong Z. Reflections on the economic evaluation of tight gas development. Financial and Economic World. 2015;**29**:23-24+154. DOI: 10.16266/j.cnki.cn11-4098/f.2015.19.020

[5] Yue T. Optimization Study on the Development Plan for Tight Gas Reservoirs in Daji Gas Field. Beijing: China University of Petroleum; 2023. DOI: 10.27643/d.cnki.gsybu.2023.001465

[6] Mingsai H, Xiang Z, Fuquan S. Research on effective development technologies for tight gas reservoirs in Xinjiang. Petrochemical Technology. 2023;**30**(03):173-175

[7] Shizhao L. Formation Conditions and Zoning of Tight gas in the Turpan-Hami Basin. China University of Petroleum (East China); 2015

[8] Chaopeng Y, Shusheng G, Guangdao L, et al. Current status and prospects of research on the seepage mechanism in tight sandstone gas reservoirs. Science Technology and Engineering. 2012;**12**(32):8606-8613

[9] Yili K, Pingya L. Current status and prospects of key engineering Technologies in the Exploration and Development of tight sandstone gas reservoirs in China. Petroleum Exploration and Development. 2007;**02**:239-245

[10] Xianyu X, Guilin B. Current status and prospects of productivity enhancement in low-permeability tight gas reservoirs. Natural Gas Industry. 1993;**02**:69-70+9

[11] Yang Z, Guanjun Z, Zhiliang S. Progress and prospects in gas well allocation theory research. Natural Gas Exploration and Development. 2013;**36**(02):44-47+86-87

[12] Xin H, Xiucheng D, Chunyue X, et al. Current status and development prospects of unconventional oil and gas exploration and development. Natural Gas and Petroleum. 2012;**30**(06):38-41+4. Available from: https://kns.cnki.net/kcms/detail/51.1183.TE.20130403.1121.011.html

Chapter 6

Perspective Chapter: Eliminating Hydraulic Fracturing by Preventing Formation Damage to Natural Gas Reservoirs while Drilling

W. James Hughes

Abstract

The current mindset in the upstream oil and gas industry is that unconventional reservoirs such as shales and tight gas sands need to be hydraulically fractured because they have very low permeability and are unable to produce naturally. If that premise were true, then why were shale reservoirs such as the Monterey in California and the Marcellus shale gas reservoir in New York so productive in the early 1900s, long before hydraulic fracturing was ever invented? The answer is simple, they were drilled without overbalanced mud systems that cause formation damage to both matrix and natural fractures that can provide darcy plus permeability. Tectonically induced natural fractures initially propagate perpendicular to the bedding plane of a formation. Over time, sedimentary beds with no dip can be tilted, thereby also tilting the natural fracture system within the formation so that even a vertical well is able to intersect a few natural fractures. Given the high dip of many formations, vertical wells are technically high-angle wells based on the definition of a horizontal well that is a wellbore drilled parallel to the bedding plane of a formation and not a wellbore drilled parallel to the surface of the earth.

Keywords: hydraulic, fracturing, eliminating, formation, damage

1. Introduction

Many early Monterey shale wells exceeded 10,000 BOPD and Marcellus shale gas production rates exceeded millions of cubic feet of gas per day without hydraulic fracturing even from wells that were only drilled vertically [1]. When most industry professionals talk about a shale reservoir being tight, they are generally referring to the matrix, which is a correct observation. What is not considered is the permeability contribution from natural micro- and macrofractures that exists in all hard (brittle) sedimentary rocks such as shales. Determination of the contribution from both matrix and natural fracture permeability is important to understanding why so-called tight rocks can produce without being hydraulically fractured. This is possible when formation damage in the reservoir has been greatly reduced or eliminated.

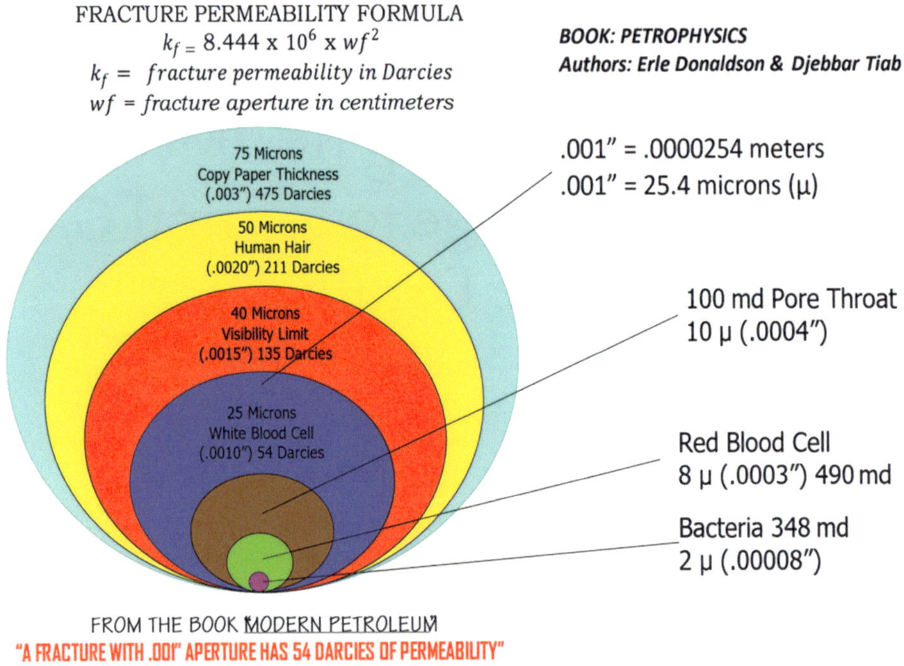

Figure 1.
Permeability chart.

It is important to remember that a single natural fracture with an aperture of 25 microns has over 50 darcies of permeability (**Figure 1**).

One of the reasons the natural fracture system is often ignored is because everyone knows the fractures penetrated by a horizontal wellbore will be plugged up with drilling mud and rendered impermeable. First and foremost, it is the job of a mud engineer to inflict maximum permeability damage upon the near wellbore region so that no hydrocarbons whatsoever enter the well while drilling. This is the reason for hydraulic fracturing, to reconnect the undamaged part of the reservoir to the wellbore. A former CEO of Halliburton has stated, "In 1947, Stanolind Oil and Gas called on Halliburton Co. to unclog a gas well in the Hugoton Basin in Grant County, Kansas." This raises the question, why clog up the natural permeability in the first place? Protecting the natural fracture permeability in shales is also ignored because our industry believes that fracing is the remedy for repairing formation damage caused by overbalanced drilling. Over many years, fracing has achieved acceptable results in conventional reservoirs with good matrix permeability so there has been little concern about formation damage in unconventional reservoirs. Formation damage prevention should be addressed because poor-quality reservoirs are unable to clean up like a good-quality reservoir. Any formation damage in a poor-quality reservoir can reduce the permeability to zero.

2. Geo-logic support for eliminating formation damage while drilling

When it comes to achieving the best recovery factor, formation damage is a well-understood and documented problem that can significantly inhibit and even prevent production of a liquid or natural gas from a reservoir. Formation damage is such a concern for the industry that the Society of Petroleum Engineers has held a conference on the subject every year for the past 50 years. Reservoir engineers refer to reduction in permeability as "skin damage." A factor for skin damage is part of every productivity formula [2]. Formation damage occurs during all phases in the recovery of hydrocarbons, which include drilling, completion, and production operations for any type of well. It is especially detrimental to reservoirs with low matrix permeability (**Figure 2**).

During the drilling of a well, damage to the permeability in the near wellbore region is intentional especially when drilling for hydrocarbons. All drilling engineers have been taught in school to use an engineered mud system to seal off 100% of the surrounding wellbore so there is no influx (referred to as a kick) of any oil or gas into the well while drilling. This requires designing a mud system with sufficient density to exert a hydrostatic weight that will be higher than anticipated bottom hole formation pressures. This is known as "overbalanced" drilling. Overbalanced drilling is acceptable for the nonreservoir formations being drilled through to reach the targeted reservoir. Once the reservoir has been encountered, it is time to switch to a drilling method designed to reduce or eliminate formation damage. That drilling method that has been around for decades is known as underbalanced drilling [3]. This means the drilling fluid is designed to have a lower density that generates a hydrostatic weight, which is lower than the formation pressure of the reservoir being drilled. Reducing the density of a drilling fluid traditionally involves the injection of a gas such as air, nitrogen, or even natural gas. Air drilling is not advisable because of the potential for downhole fires. The preferred gas for lowering the density of the drilling fluid is nitrogen. Now that the industry has membrane units to sieve and remove the majority of oxygen from air, nitrogen, which is inert, can be used to dilute the drilling fluid. In the early days before membrane technology, reservoirs drilled underbalanced used cryogenic nitrogen, which was extremely expensive depending on the location of the cryogenic plant [4].

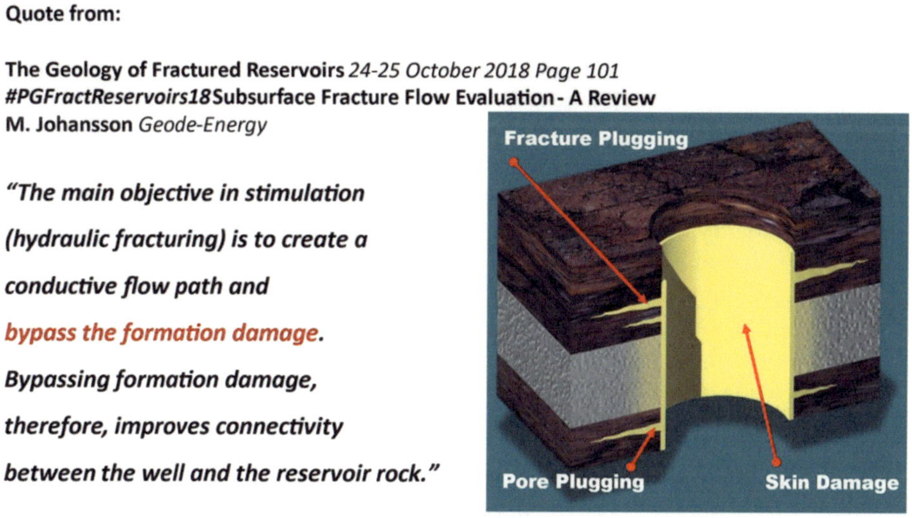

Figure 2.
Formation damage.

A relatively new and refined overbalanced drilling method known as "Managed Pressure Drilling" (MPD) [5] uses underbalanced equipment to drill slightly overbalanced. The method is appropriate for the nonreservoir sections of a well. The mud system deployed in MPD is designed to be just slightly overbalanced. In other words, drilling engineers still do not want to experience any "kicks" from oil or gas entering the wellbore during drilling but do want to be closer to "at-balance" to primarily improve penetration rates. This is accomplished by reducing the "chip-hold-down-effect" that slows penetration rates. The higher the mud density and resulting hydrostatic weight, the more difficult it is to remove drill cuttings from beneath the bit. Cuttings held in place by heavy mud are then drilled twice.

The trade-off to achieving higher penetration rates by drilling slightly overbalanced means the well is more likely to see some influx of oil and gas, especially while making a connection. During this process of adding drill pipe, the wellbore will lose back pressure from friction when the mud pumps are turned off. This loss of back pressure will lower the effective hydrostatic pressure against the formation to a point that may be below the pressure within the formation, thereby allowing oil or gas to enter the wellbore. This is why MPD operations may use a separate pump to increase back pressure on the annulus to replace the loss of friction pressure from circulation. To allow the kick to be contained and circulated out of the well, managed pressure drilling requires a vital piece of equipment that was developed for underbalanced drilling known as a "rotating control device" (RCD) [6]. This piece of equipment is similar to an annular BOP except that it includes a bearing that allows the single piece sealing element to rotate with the drill pipe. An annular BOP is designed for static (nondrilling) well control and an RCD is designed for dynamic (rotating drill pipe) well control.

Underbalanced drilling (UBD) is different than managed pressure drilling because UBD does encourage influx of hydrocarbons while drilling. This drilling technique is known in the industry for reducing formation damage from plugging caused by overbalanced mud systems. The technique was originally developed not to improve productivity but to reduce nonproductive time, thereby improving drilling performance in the nonreservoir section of a vertical well. After horizontal drilling became commercial, the drilling method was applied in conventional reservoirs to improve productivity with great success as long as the well remained underbalanced 100% of the time. UBD has been applied primarily in traditional reservoirs, aka reservoirs with low capillary pressures where only one formation damage mechanism needs to be addressed. It has not been used in shale formations because traditional underbalanced techniques do not address all of the damage mechanisms in shale including staying underbalanced 100% of the time even while rigging down.

With the industry now focused on unconventional reservoirs, drilling and completion engineers should be pursuing technology that eliminates all types of formation damage mechanisms that are common to shales. There should be no higher priority than protecting permeability to increase recovery factors. It is no longer good enough to just prevent plugging caused by overbalanced mud systems because there are several other damage mechanisms inherent to poor-quality reservoirs. In other words, if the goal is to eliminate hydraulic fracturing, then all types of damage must be addressed.

The reason this is so important is because of natural fracture permeability that exists in all rocks [7] especially in low matrix permeability rocks. The first step in determining if a reservoir can produce naturally is to determine "collective permeability." Collective permeability is the summation of matrix permeability with the multiplication of fracture porosity and micro–/macrofracture permeability. Once it is determined for a so-called tight formation, engineers will see the well is capable of producing naturally, without stimulation since the reservoir will be drilled horizontally and perpendicular to the primary stress direction where it will encounter the maximum number of natural fractures (**Figures 3** and **4**).

The software required to understand "collective" permeability in a dual permeability system is called Discrete Natural Fracture Modeling (DNFM) [8]. The computer-generated models show 10s to 100 s of thousands of natural fractures over very short horizontal distances, which can now be intersected with a properly oriented wellbore. If the near wellbore fractures are protected from formation damage, this would explain why unconventional reservoirs can produce naturally without

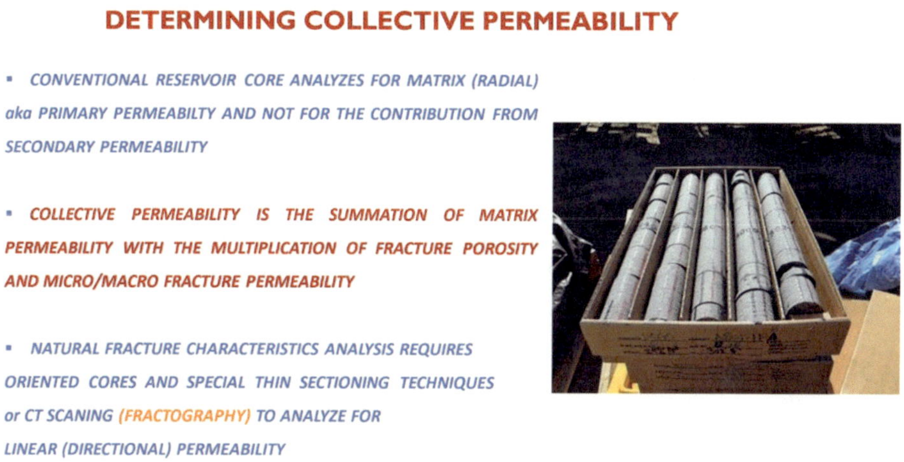

Figure 3.
Core.

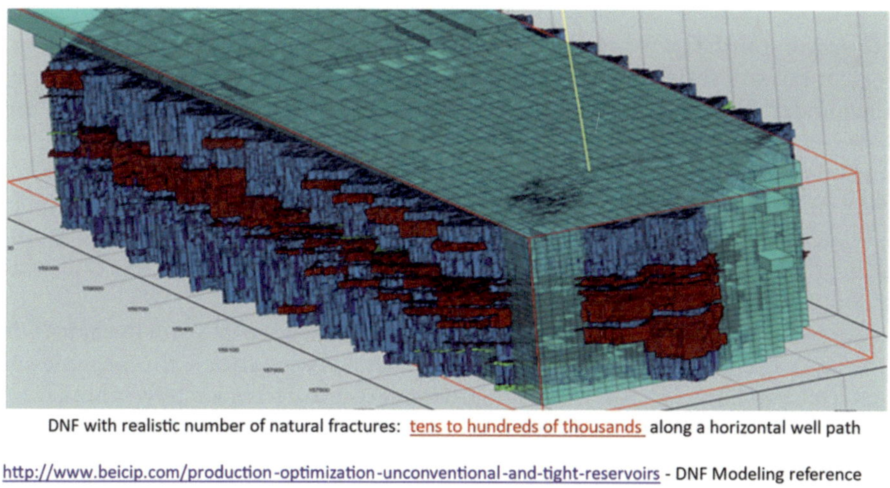

Figure 4.
Discrete natural fracture modeling.

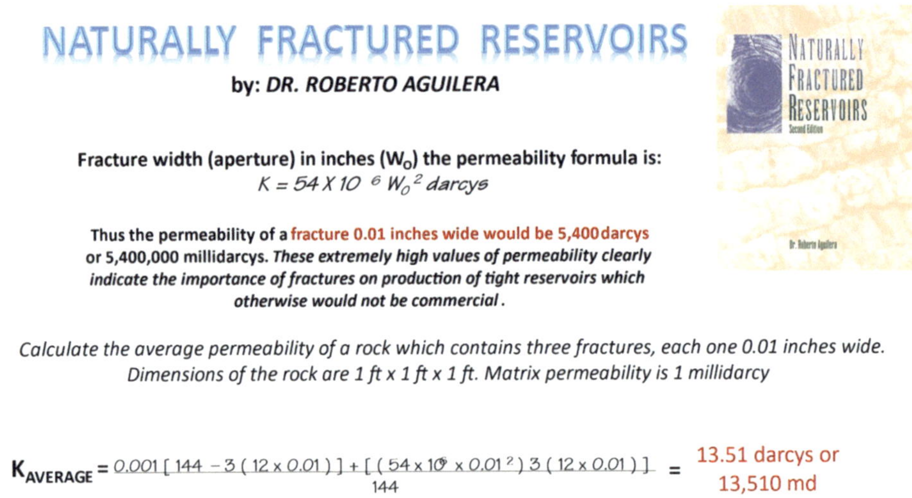

Figure 5.
Collective permeability.

hydraulic fracturing. For example, the Spraberry formation in West Texas is described as a tight reservoir with permeability on the order of one-tenth of a millidarcy. It is also known to be naturally fractured having two distinct fracture sets. It has been determined using DNFM software that the collective permeability of the Spraberry in one direction is actually 30 millidarcies, which is sufficient to produce naturally, if all formation damage mechanisms are eliminated.

Dr. Roberto Aguilera provides a good example of "collective" permeability, which he calls "average" in his book NATURALLY FRACTURED RESERVOIRS (**Figure 5**).

3. Formation damage mechanisms in hydrocarbon bearing shales

Reservoirs with low matrix permeability have high capillary pressures meaning a reservoir can still suffer permeability damage from imbibition even if the wellbore is drilled underbalanced [9]. This happens when capillary (suction) forces exceed the underbalanced condition. It is common for an underbalanced wellbore to be drilled with 500 psi less pressure than the formation pressure. If capillary pressures are higher, imbibition will still occur. Imbibition is preventable. Most shale formations are "water wet." Thus, the solution to stop imbibition in shales is to drill with a non-wetting fluid such as mineral oil. Imbibed water that becomes irreducible can reduce matrix and microfracture permeability to zero [10].

Fines migration [11] is another documented damage mechanism in shales with high clay content. During traditional underbalanced operations, this becomes a real problem because UBD wells are often overproduced to keep the annular wellhead pressure from getting too high while drilling. Over producing the well by opening the choke to reduce the pressure on the rotating control device (RCD) causes fines to mobilize and then bridge off at the wellbore. This problem is well understood in production operations where wells eventually plug off from the mobilization of fines formation when wells are over produced. If a portion of the formation fines are produced back to surface, the fines are called "basic-sediment."

Figure 6.
UB vs. OB.

This plugging that results from solids moving through the reservoir to the wellbore is similar to plugging from solids in overbalanced drilling muds that are injected into the reservoir.

Permeability damage from friction-reducing polymers aka gel damage can be significant. Friction-reducing additives are used to make "slick" water for hydraulic fracturing. They can also be used in the drilling of horizontal wellbores to reduce friction between rotating drill pipe and the wellbore wall especially in long laterals. Breakers are not used at the end of drilling to reduce the viscosity of the gel because wells are generally not completed open hole. After drilling, a liner is usually run and cemented, which causes even more damage so why worry about gel damage.

Corrosion inhibitors are one of the biggest contributors to permeability reduction in a reservoir because these additives cause wettability alterations. In traditional underbalanced drilling, it is common to use an inhibitor to protect drill pipe from corrosion. Using nitrogen to reduce the density of the drilling fluid, which is usually water, does not mean there is no oxygen being pumped downhole. Nitrogen membrane units typically cannot separate all of the oxygen from the nitrogen being fed into the compressors. The mixture is about 95% N2 and 5% O2. Even 5% O2 at 3000 psi being injected into the stand pipe causes severe corrosion to the drill string, so much so "corrosion coupons" are run in a drill pipe connection and the drill pipe rental company can bill the customer for loss of "remaining body wall" thickness (**Figure 6**).

4. Formation damage from hydraulic fracturing

Hydraulic fracturing is a major contributor to severe formation damage [12] even though it is referred to as a stimulation technology. If a mud engineer has done his or her job to completely seal off (zero permeability) the near wellbore region while drilling and the completion engineer has cemented a liner in place resulting in zero or near zero production after perforating, then the well must be hydraulically fractured to reconnect the undamaged part of the reservoir to the wellbore. If production is then obtained after hydraulic fracturing, then it must be a stimulation technique even if 90% of the hydrocarbons are left in the ground. To many people, a 10% or less recovery factor is unacceptable [13]. This poor average recovery factor is in part due to formation damage from hydraulic fracturing fluids and additives (**Figure 7**).

Experts on this subject have identified numerous formation damage mechanisms caused by hydraulic fracturing especially in shale reservoirs. Shales can be 35% by volume composed of clay minerals. These clay minerals such as smectite and illite react adversely when they come into contact with water and yet millions and millions of gallons of water are used to hydraulically fracture shale wells. Drillers have known for decades the issues that can occur when water-based drilling fluids come into contact with shales that contain clay minerals. Injection of water into shale formations can significantly have a negative impact on recovery factors because the formation damage mechanism known as "water block" [14] dramatically reduces permeability. Shales are generally undersaturated with respect to water. This means they are like a sponge and can imbibe more water. In shale reservoirs where matrix permeability is low, capillary pressures are high. Thus, when water is imbibed it becomes bound to water, thereby creating a blockage that reduces permeability [15]. It can be argued that shale wells are never truly depleted of hydrocarbons. Production drops making it look like depletion, but in reality, permeability has been reduced by formation damage mechanisms and thus making the well appear depleted.

Another significant formation damage mechanism from hydraulic fracturing is caused by the use of friction reducers and referred to as "gel damage." As mentioned earlier, friction reducers are used to create slick water for transporting frac sand. This friction-reducing gel is so potentially damaging to the reservoir that the Society of Petroleum Engineers featured an article of the subject called "Gummy Bears" [16].

Figure 7.
Plugging from gel.

5. The next generation of underbalanced drilling technology eliminates all formation damage mechanisms in shale gas reservoirs

The next generation of underbalanced drilling technology and methods described herein requires no research. Each technology is based on proven equipment that will simply be repurposed to prevent all types of formation damage mechanisms so the "protecting the reservoir" drilling method can be used as a completion technique in shale gas formations. This begins with using a jet pump, which is a well-proven artificial lift technology to induce the underbalanced condition (see drawings in US Patent No. 11,168,526) by pumping the well while drilling to lower the fluid level in a well and thus reducing hydrostatic weight to a point that is slightly lower (near balanced) than the pressure in the reservoir. After all, the primary objective of artificial lift equipment, including jet pumps, is to eliminate back pressure on the producing formation (**Figure 8**).

This technique to create the underbalanced condition eliminates the need for nitrogen membrane and compression equipment used in traditional UBD and thus reduces costs by 90%. The jet pump can be placed in a near-bit stabilizer and uses the bit as the flow path into the jet pump. This is similar to reverse-circulation except in this situation only oil, gas, and drill cuttings flow into the jet pump. Reservoir fluids being produced through the bit will cool the bit along with the significant pressure drop caused by jet pumps. Jet pumps are a well-proven technology and are famous for being able to pump anything that includes large rocks during dredging operations. Plus, jet pumps have no moving parts, which is an important feature for a tool that is deployed downhole. For decades, jet pumps have been used to vacuum horizontal wells after they have been hydraulically fractured to remove frac sand, drilling mud, and slick water from the reservoir. Quantum Downhole Systems [17] has a nice animation on their website that will show a jet pump with a jetting head on the end of concentric coiled tubing. The design shown in US Patent 11,168,526 (see drawings) is different from the quantum system because the jet pump with a drill bit can be run on jointed pipe and allows no drilling fluid to exit through the drill bit like the quantum jetting head.

Another important benefit from using a jet pump to induce the underbalanced condition is that it ensures the underbalanced condition is maintained 100% of the time, which is critical to the success of eliminating formation damage and completion of the well. The area where damage still occurs when using traditional underbalanced

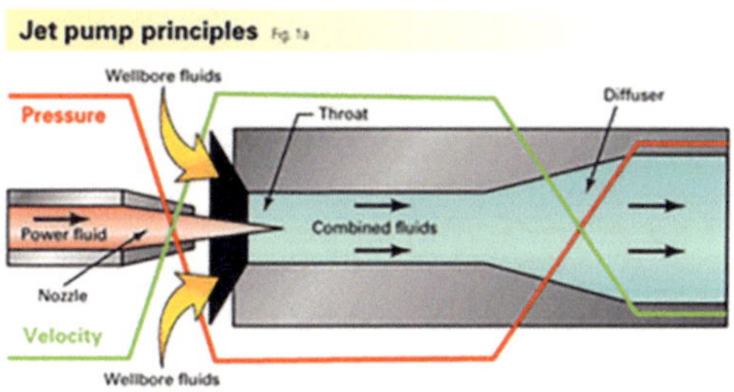

Figure 8.
Jet pump principles. Image from: https://www.ogj.com/general-interest/companies/article/17233947/coiled-tubing-vacuum-removes-drilling-induced-damage.

methods is in front of the bit. This is where most of the damage occurs because the area in front of the bit is highly overbalanced until the fluid makes a U-turn and heads back uphole [18]. This is why traditional underbalanced drilling does not work in shales. This instant injection of water, nitrogen, and oxygen causes severe damage to the near wellbore region known as the damaged zone. The use of the jet pump solves the problem by using the nozzles in the drill bit as flow paths for reservoir fluids and drill cuttings to enter the jet pump and then pump back to the surface. The picture in the upper right-hand corner is from a video on U-Tube titled "Directional Drilling Nightmare" [19]. Just imagine the damage being done to the reservoir ahead and around the bit when drilling occurs downhole (**Figures 9** and **10**).

The next generation of underbalanced drilling needs to improve upon safety at the rig floor so people will feel safe while a well is flowing gas and/or oil during drilling operations. Improving safety begins with a new rotating control device that can be placed below the BOP. All RCDs today are placed on top of the BOP stack, which means high pressure if being controlled very near the rig floor, which makes floor hands nervous. They know the only thing between them and a blowout is a small sealing element that is only about one foot tall. The traditional RCD is positioned on top of the BOP stack because the "single piece" sealing element can only be replaced through the top of the RCD. What is needed is an RCD with a "two-piece" sealing element (see drawings in US Patent No. 11,441,383) that allows the sealing elements to be removed through side doors just like a pipe ram. This type of RCD with side doors can be placed on top of the wellhead and below the BOP stack. This BOP barrier will be many feet away from the rig floor and will allow the traditional RCD to be used as a backup barrier, which will greatly improve for safe operations.

The next critically important piece of equipment to further enhance safety of UB operations is a mechanically operated subsurface safety valve to create a downhole lubricator to eliminate the costly and dangerous snubbing operation. (See drawings in US Patent Application No. 17/709,210.) A subsurface safety valve will be actuated by a rotating wellhead that can rotate a tieback liner that opens and closes the safety valve. In other words, the tieback liner serves as a mechanical arm because it is uncemented. As mentioned earlier, it is imperative to stay underbalanced 100% of the time when

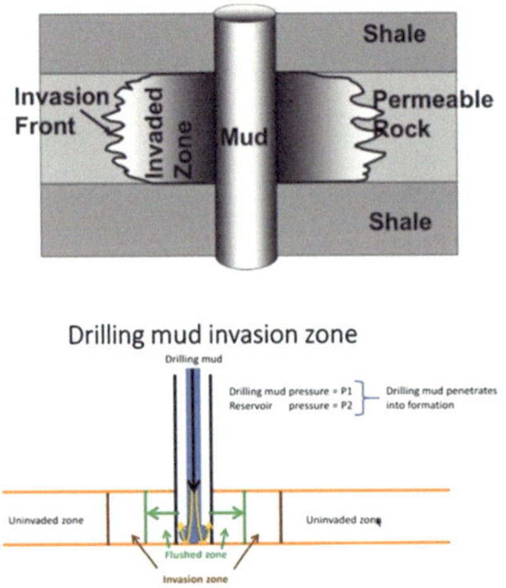

Figure 9.
Mud invasion.

Figure 10.
Hydraulic forces exiting drill bit.

drilling the reservoir. Therefore "killing" a well to trip a bottom hole assembly in or out of a well is not acceptable. Snubbing units can be used but are dangerous and expensive because they sit on standby 90% of the time. Thus, the solution is to "lubricate" the bottom hole assembly in or out of the well by using the drill pipe ram or annular at the surface and a subsurface safety valve to create lubricator, which is below the surface pressure isolated chamber. This method of running tools is well proven and the "lubricator" is a permanent installation that can be used later during production operations to run artificial lift equipment in and out of a well under pressure without killing the well or having to snub.

The next generation of underbalanced operations should provide a means of well construction to change the flow path for hydrocarbons returning to surface and into production separation equipment. (See drawings in US Patent No. 11,255,144.) The well plan will consist of drilling the vertical well and curve overbalanced to the reservoir. Production casing will be set and cemented from surface to the reservoir and hung in the intermediate casing at a depth where the subsurface safety valve is located. The SSSV will be run in on a tieback liner that has been perforated above the tieback seal assembly and below the safety valve before running in the hole. The subsurface safety valve is then run in the hole and the tieback liner is hung in the rotating well head. Rotating wellheads have been used in artificial lift operations for decades to prevent uneven tubing wear caused by sucker rods. In this application, rotating the tieback will advance or retract a threaded actuator that will open and close a subsurface flapper valve.

The casing program that allows for the installation of a downhole lubricator automatically creates a second annulus between the tieback liner and the intermediate casing creating an alternate pathway back to intermediate casing wellhead that can be located in a cellar. A tubing flow line is then connected to the outlet flange of the wellhead and run a couple of feet underground to the production separation equipment. For obvious reasons, this is a major safety improvement for underbalanced operations having no high-pressure flow lines lying on the surface.

Once the BOP equipment is in place, the well is now set up to drill the reservoir underbalanced. With the lower RCD fully engaged to seal off the annulus between the production casing and drill pipe, the outer annulus provides the only route for hydrocarbons to return to the surface by flowing through the perforations just below the SSSV and into the outer annulus. This next-generation design for near-balanced reservoir drilling operations has moved pressure and flow as far away from people as possible.

6. Rethink how a reservoir should be drilled

Preventing formation damage while drilling is also critically important when injecting gases or fluids into a formation for CO2 sequestration in order to make natural gas production carbon neutral. The same "protecting the reservoir" logic applies whether producing from or injecting into a rock formation because a sealed off wellbore prevents flow in either direction. The benefits from eliminating formation damage support the effort to make natural gas a transition fuel. The desire for natural gas to be environmentally friendly has led to a new term emerging within the industry, "green natural gas." It can have multiple meanings that include natural gas that is not produced from fraced wells, natural gas whose CO2 by-product is sequestered, or natural gas (methane) that is produced from biomass. For example, if an LNG exporter purchases natural gas from wells that have not been fraced, then they can claim their gas is "green" aka "frac-free natural gas." If a natural gas power plant sequesters their CO2 emissions, then the utility can claim the natural gas used to generate electrical power is "green" because their power plant is carbon neutral. Protecting the disposal reservoir provides an environmental solution for power plants because sequestering wells can be drilled on the same location as a natural gas power plant, thus eliminating the need for a pipeline to sequester CO2 at a remote location. Another application for "green natural gas" is the generation of hydrogen. Natural gas is a key component in the hydrogen manufacturing process. If the CO2 by-product is sequestered, then the new term for carbon neutral hydrogen is "blue hydrogen." What if a drilling technology is able to deliver "cyan hydrogen" (blue-green), which is hydrogen whose CO2 by-product is sequestered (blue) for a carbon neutral footprint and the natural gas used to make the hydrogen is frac-free (green) natural gas. This shale gas development approach can make projects carbon neutral by simply drilling a CO2 sequestration well using the same nondamaging drilling technology that is used to eliminate hydraulic fracturing.

7. Conclusion

The origin of formation damage began 100 years ago, and its cause was attributed to the original Hughes Tool Company. The company's roller cone bit design forever changed how the oil and gas were drilled. Before rotary drilling, wells were drilled with cable tools, which mechanically removed cuttings from the wellbore. To remove cuttings from a well being drilled with, the Hughes roller cone bit required the circulation of drilling fluid (mud) down the drill pipe through the bit returning to the surface via the annulus. This was a major change that unknowingly at the time caused damage to the near wellbore permeability of a reservoir. Formation damage caused by overbalanced drilling as it is known today eventually led to the development of hydraulic fracturing, a completion technique that attempts to undo the damage. The next generation of underbalanced drilling technology will change how the reservoir is drilled so that all types of formation can be eliminated even in shales. After all, in our day and time, we now have the ability to drill horizontally into the reservoir. This means an operator can switch from drilling the nonreservoir vertical section and curve of the well using conventional overbalanced drilling methods to using new technology and procedures to drill slightly underbalanced into the reservoir. This logical change to a conventional drilling program will eliminate the need for hydraulic fracturing, which can reduce total well costs by 50%.

Author details

W. James Hughes
Hughes Tool Company, Denver, Colorado, USA

*Address all correspondence to: nbrd.jh@gmail.com

IntechOpen

© 2024 The Author(s). Licensee IntechOpen. This chapter is distributed under the terms of the Creative Commons Attribution License (http://creativecommons.org/licenses/by/4.0), which permits unrestricted use, distribution, and reproduction in any medium, provided the original work is properly cited.

References

[1] SLB. The Defining Series: Formation Damage. 2016. Available From: https://www.slb.com/resource-library/oilfield-review/defining-series/defining-formation-damage

[2] Journal of Australian Energy Producers. Evolution of Damage Mechanisms and Skin Factor in Tight Gas Reservoirs. 2011. Available From: https://www.publish.csiro.au/aj/aj10045

[3] PetroWiki. PEH: Underbalanced Drilling. 2017. Available From: https://petrowiki.spe.org/PEH:Underbalanced_Drilling#:~:text=Metric%20Conversion%20Factors-,What%20is%20Underbalanced%20Drilling%3F,formations%20that%20are%20being%20drilled

[4] United Nations University. Drilling Practice with Aerated Drilling Fluid. 2008. Available From: https://rafhladan.is/bitstream/handle/10802/6674/UNU-GTP-2008-11.pdf?sequence=1

[5] PetroWiki. SPE Managed Pressure Drilling. 2020 Available From: https://petrowiki.spe.org/Managed_pressure_drilling#:~:text=Managed%20Pressure%20Drilling%20(MPD)%20is,pressure%20profile%20throughout%20the%20wellbore

[6] PetroWiki. Surface Equipment for UBD Operations. 2015. Available From: https://petrowiki.spe.org/Surface_equipment_for_UBD_operations

[7] GEOxPro. Fracture, Fracture Everywhere-Part I. 2014. Available From: https://geoexpro.com/fracture-fracture-everywhere-part-i/

[8] Journal of Petroleum Technology. Discrete-Fracture and Dual Permeability Models for Multiphase Flow in Naturally Fractured Reservoirs. 2011. Available From: https://onepetro.org/JPT/article-abstract/63/07/84/199851/Discrete-Fracture-and-Dual-Permeability-Models-for?redirectedFrom=fulltext

[9] Journal of Natural gas Science and Engineering. Review on Formation damage Mechanisms and Processes in Shale Gas Reservoir: Known and to be Known. 2016. Available From: https://www.sciencedirect.com/science/article/abs/pii/S1875510016302153

[10] Journal of Natural Gas Science and Engineering. Permeability Evolution of Shale during Spontaneous Imbibition. 2017. Available From: https://www.osti.gov/servlets/purl/1343057#:~:text=the%20permeability%20reduction%20due%20to,clay%20content%20or%20minor%20swelling

[11] PetroWiki. Formation Damage from Fines Migration. 2015. Available From: https://petrowiki.spe.org/Formation_damage_from_fines_migration

[12] SPE Eastern Regional Meeting. The Role of Formation Damage in Hydraulic Fracturing Shale Gas Wells. 2011. Available From: https://onepetro.org/SPEERM/proceedings-abstract/11ERM/All-11ERM/SPE-148778-MS/150543

[13] Search and Discovery. Improving Unconventional Hydrocarbon Recovery by Reducing Formation Damage. 2016. Available From: https://www.searchanddiscovery.com/pdfz/documents/2016/80562cercone/ndx_cercone.pdf.html

[14] SPE European Formation Damage Conference. Water Induced Formation

Damage in Unconventional Reservoirs. 2009. Available From: https://onepetro.org/SPEEFDC/proceedings-abstract/09EFDC/All-09EFDC/SPE-121944-MS/146372

[15] ResearchGate. Factors Affecting Water Blocking and Gas Flow from Hydraulically Fractured Gas Wells. 2013. Available From: https://www.researchgate.net/publication/238057208_Factors_Affecting_Water_Blocking_and_Gas_Flow_From_Hydraulically_Fractured_Gas_Wells

[16] Journal of Petroleum Technology. Solving the Gummy Bears Mystery May Unlock Greater Shale production. 2020. Available From: https://jpt.spe.org/solving-gummy-bears-mystery-may-unlock-greater-shale-production

[17] Quantum Downhole Systems. JetVac System. 2023. Available From: https://www.quantumdownhole.com/

[18] Science Direct. Effect of Overbalance Pressure on Formation Damage. 1992. Available From: https://www.sciencedirect.com/science/article/pii/S0920410502002681

[19] YouTube. Oilfield Directional Drilling Nightmare. 2011. Available From: https://www.youtube.com/watch?v=8CmsG4Z_C9c&t=100s

Chapter 7

Shale Gas: Environmental and Radiological Aspects

Beatriz Hernandez-Mendez, Sergio Suarez-Contreras, Juan Carlos Sanchez-Meza, Edith E. Gutierrez-Segura and Miguel Balcazar

Abstract

Radiological effects on population and environment due to fracking unconventional hydrocarbons is given in this study. Evaluation on environmental effects concerning extraction process is given. Effects on fauna and humans during fracking process, including on air, water and soil, are mentioned. Unconventional oil and gas extraction generates Natural Occurring Radioactive Materials (NORM); their associations to radiological aspects are comment. Our environmental radiological baseline studies at three prospecting fracking zones in Mexico showed no radiological hazard to population. The natural gamma radioactivity before fracking had a maximum equivalent dose rate of 1.83 mSv a^{-1}, in the low value of the acceptable background interval (1–13) mSv a^{-1} stated by UNSCEAR. The 226Ra and 228Ra maximum concentration obtained in soil (59.2 ± 3.6 and 38.5 ± 5.0 Bq kg^{-1}, respectively) for this baseline are both in the average global-interval concentration reported by UNSCEAR; gross alpha and gross beta in water were all below the suggested limits by the World Health Organization of 0.5 and 1.0 Bq L^{-1}, respectively, for drinking water. An important conclusion is that water stress is an important issue before considering fracking activities in Mexico.

Keywords: shale gas, NORM, natural radiation, fracking, environmental baseline

1. Introduction

An alternative exploitation of oil fields was introduced in the twenty-first century, Section 2 provides the global situation based on unconventional hydrocarbons; for example, United States of America began the exploitation of using fracking technique; similarly, other countries evaluate their reserves of unconventional hydrocarbons, fracking process which is briefly described in Section 3. Associated with fracking technique environmental issues are addressed in Section 4. Section 5 deals with the exposure to ionizing radiation by natural radionuclides present in the earth's crust from the decay chains of ^{238}U and ^{232}Th, which can modify their natural concentrations due to the drilling and exploitation of unconventional hydrocarbon wells. The increase in natural radionuclide concentration (NORM) is a study topic associated with not only oil exploitation, but also mineral extraction industry; some countries, such as Mexico, lack of having reference values (baselines) for the concentration of natural radionuclides in air, water and soil. Section 6 reports the first radiological baseline at three zones in Mexico, identified as unconventional oil resources, prior to exploitation. The baseline includes the equivalent dose by gamma emitters from natural radionuclide present in soil, which determines the population exposure rate to ionizing radiation from natural radionuclide and the environmental radiological effect. Radionuclide content from ^{238}U and ^{232}Th decay series was also determined in soil, as well as alpha and beta emitters from water samples. Comments are also on limitation and fragility of groundwater in Mexico. The International Atomic Energy Agency supported this baseline determination, prior to a possible development of fracking in Mexico.

2. Unconventional hydrocarbons: Global situation

2.1 Oil and gas production

According to the International Energy Agency (AIEA), natural gas production in the world was 3274 billions of cubic meters (Bcm) in 2010 and 4138 Bcm in 2022, with an unconventional gas contribution of 15.4 and 30.6%, respectively. Production forecast for 2030 would be 4299 Bcm having 32.6% of unconventional natural gas and for 2050 is expected to be 4173 Bcm with 27.7% of unconventional gas contribution [1].

The maximum rate at production peak of any natural resource defines its disponibility declination. In the case of oil, peak production occurs when approximately half of the recoverable oil was extracted from the reservoir. In fact, most oil-producing countries, including Indonesia, United Kingdom, Norway, and United States of America, have already passed peak crude oil production several years or decades ago. However, their drop production has been balanced by discoveries and production growth in other parts of the world, as well as additional development of unconventional crude oil and natural gas resources found in tight and shale formations [1]. After the peak production, follows a stage where it is increasingly expensive to produce resources [2].

Countries are moving toward an energy transition, consisting in changing production models and energy consumption to reduce greenhouse gas emissions. This implementation toward clean energy is established in the international Paris Agreement by more than 180 countries, as part of the United Nations Framework Convention on Climate Change (UNFCCC). The main objective of the agreement is to limit the increase in the global average temperature below 2°C, encouraging the use of low-carbon energy sources to reduce greenhouse gas emissions [3].

However, today, 80% of the energy use still comes from fossil fuels, whereas for electricity generation 70% of the total is from coal, oil and gas [4]. Natural gas is an alternative for replacing coal toward renewable energy sources but it keeps dependence on fossil fuels. Measurements by the United Nations Environment Programme (UNEP) show that methane emissions from gas operations are much higher than previously estimated, but if management is adequate, those emissions are not as problematic in terms of global warming as coal and oil [5].

2.2 Unconventional oil and gas resources

The demand and depletion of conventional oil resources are key factors at the global level to consider unconventional resources as an alternative for expanding the fossil fuel market for some countries. The experience of the United States of America in fracking exploitation is undoubtedly an important precedent of energy autonomy based on this oil extraction technique. Global research on unconventional hydrocarbon resources to conventional hydrocarbon studies is the ratio 8:2, and this ratio is mainly in shale gas, since for shale oil, conventional oil resources are equivalent to unconventional oil [6]. The United States of America has increased shale gas production since 2007; shale gas is currently 77% of its total dry natural gas production and projections show a 22.5% growth production from 2022 to 2050 [7–9].

North America, Central America, South America, Europe and Asia Pacific have unconventional resources in a proportion of 60.9 to 72.8% of their total natural gas resources; in a minor percentage are the Middle East, Eurasian and African regions. The United States of America is the largest shale gas producer, but China is the country with the largest reserves [1, 10]. **Table 1** shows the leading countries of unconventional gas and crude oil resources as stated by US Energy Information Administration [10].

The main shale gas and oil production is in the United States of America, China, Canada and Argentina [11]. The fracking practice in Canada began in 2005; tight gas contributes to 65% and shale gas 3% from the total production of natural gas. In Argentina, shale gas contributed to 48% of total national production in 2021. In China, production in 2020 was 53 billions cubic feet per day (Bcfd).

2.3 Unconventional resources: explotation effects

According to the National Institute of Environmental Health Science after more than two decades of unconventional resources exploitation in United States of America, research focuses on determining the effect and extent effects to human health and the related costs [12]. The United States Environmental Protection Agency

Country	Shale Gas Wet trillion cubic feet (Tcf)	Country	Tight Oil billion barrels (Bb)
China	1115.2	USA	78.2
Argentina	801.5	Russia	74.6
Algeria	706.9	China	32.2
USA	622.5	Argentina	27.0
Canada	572.9	Libya	26.1
Mexico	545.2	United Arab Emirates	22.6
Australia	429.3	Chad	16.2
South Africa	389.7	Australia	15.6
Russia	284.5	Venezuela	13.4
Brazil	244.9	Mexico	13.1

Table 1.
Countries having shale gas and oil gas reserves. Adapted from United States Energy Information Administration. World Shale Resource Assessments. 2015.

found that hydraulic fracturing activities could affect drinking water resources under certain circumstances and non-direct evidence that water pollution related to hydraulic fracturing influences human health [13]. Regarding air pollution research up to 2022, it is suggested that pollutants from unconventional oil and gas exploration are associated with adverse health effects in elder people, who live facing airstream from production centers, increasing their risk of mortality [13]. Environmental and human health effects have to be taken into account by applying the best international practices and developing an appropriate regulatory framework. Shale gas production in North America over the past decade has contributed to one-third of the total increase emissions from all sources globally [14]. Also to the exploitation process transport natural radionuclides from the source rock to the surface, increasing natural-radionuclide concentration from ^{238}U and ^{232}Th decay chains, this radiological aspect was addressed in detail for the Marcellus deposit in the United States of America [15, 16].

3. Unconventional hydrocarbons and fracking

3.1 Origin of oil and gas

Crude oil is the product decomposition of organic material exposure to high pressures and temperatures during certain time [17]. Oil and gas are formed in the so-called "source rock." Subsequently, hydrocarbons migrate into a porous rock named "reservoir rock" where extraction takes place. Oil and/or gas remain confined in the reservoir rock if a structural or stratigraphic "trap" avoids migration. In addition, there is an impermeable rock known as "seal rock," which prevents hydrocarbon migration to the surface.

An oil system to generate and store hydrocarbons is the geological components that include a reservoir rock, a migration path, a reservoir rock and a seal rock [18, 19]. Oil and gas survey and potential areas are around basins or regions where there is a certain probability of the existence of an oil system.

Conventional deposits are reservoirs where gas and oil accumulate by migration from the source rock to high porosity and permeability formations; these conventional deposits are exploited using traditional techniques since the second half of the nineteenth century. The reservoirs that are still in the source rock are unconventional reservoirs, characterized by their low permeability and porosity, and therefore to obtain the hydrocarbon trapped inside, fracking techniques are required. These resources started exploitation since the second decade of the twenty-first century [20].

3.2 Exploitation of unconventional resources

Conventional resources produces economical amounts of oil and gas from vertical wells with standard techniques and minimal methodical stimulation, while unconventional resources require special reservoir engineering processes such as multidirectional horizontal drilling and hydraulic fracturing to produce profitable amounts of hydrocarbons [21]. Reservoirs are classified into conventional and unconventional, based on the value of their permeability [22]; thus, a conventional deposit is suitable if its permeability is greater than 0.1 milliDarcy (mD), while smaller permeability is taken as unconventional.

Figure 1 shows a distribution diagram of unconventional and conventional gas resources. Unconventional hydrocarbons such as tight gas, tight oil, tar sands and

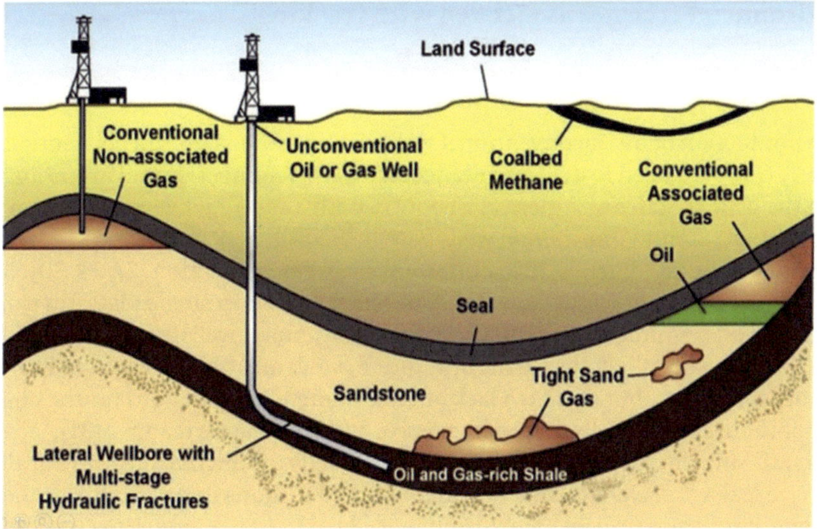

Figure 1.
Conventional and unconventional reservoirs. Modified from U.S. Energy Information Administration, August 2016. Annual Energy Outlook 2016 with Projections to 2040: Report DOE/EIA-0383, p 256.

heavy and extra-heavy oils originate in reservoir rocks. Likewise, some unconventional hydrocarbons are in its formation rock such as shale gas, shale oil, methane hydrates and gas in coal deposits. Tight gas, tight oil, shale gas and shale oil unconventional resources are extracted by fracking using vertical hydraulic-fracturing technique combined with multidirectional horizontal drilling [17, 20, 23, 24].

The fracking technique to extract unconventional resources in the shale bedrock is a combination of multidirectional horizontal drilling with high-volume hydraulic fracturing. The technique is applied on platforms where vertical drilling is carried out to meet the target formation, which can be from 1000 to 1500 m in depth. At different depths, the drilling pipe is removed, and a cement casing steel pipe is introduced to create barriers between the well and its exterior, reducing the possible aquifer contamination that could present across the drilling trajectory. Subsequently, the drilling angle is changed to a lateral wellbore with multi-stage hydraulic fractures that sometimes go along to 2000 m, in order to expose the largest oil and gas-rich shale area (see **Figure 1**). Then, the pipe is punched to create cracks to inject fluids at high pressure based on a mixture of water, sand and chemical additives. Here, the reservoir is hydraulically stimulated allowing the cracks to expand, and the sand acts as a proppanting agent to support the cracks and prevent them from collapsing. Fracturing is performed at different horizontal stages; at the end of each fracturing stage, a plug is introduced to contain the injected fluid and the hydrocarbon. Then, fracturing continues in some cases up to 30 horizontal stages. Once the stages concluded, all plugs are removed to allow the fluids upcoming to the surface. The output fluid at the beginning consists mostly of injected water followed by considerable fraction of water from the reservoir, which both must be treated before use or storage; then, the production hydrocarbons start to flow [20, 23, 25].

The fracking practice is controversial in environmental aspects; the most important risks of impact refer to high consumption of fresh water and the risk contamination of surface groundwater, emissions of pollutants into the air and greenhouse gas emissions that contribute to climate change [26, 27].

4. Environmental issues associated with fracking

4.1 Fracking impact

The production of an unconventional well is low as compared to a conventional one; therefore, there is a need to drill simultaneously a large number of unconventional wells [28]. In the United States of America, a country with a sustained development of shale gas production, around one million wells have been drilled up to 2022 on 11 formations contributing to approximately 70% of its total shale gas production [27, 29, 30]. Massive production is important environmental issue, since wells on a single platform would not satisfy the need resource of a country; many platforms and wells are needed that would increase the "footprint" on the earth. In addition, wells and platforms should be within regulatory tolerances, but there is a lack of environmental regulations for the possible impact of thousands of wells drilled in massive and cumulative manner [31].

A significant situation is that throughout the wells productive life some of them could have leakage accidents, due to unexpected conditions in this high-risk industry, producing significant impact on environment [32, 33]. Some references focus on evaluating wells and platforms conclude a positive process evaluation with low environmental risk, while those that mark a high risk are based on the cumulative effects in an area, whose scientific evidence is not yet available.

The exploitation of shale oil and shale gas by fracking involves several environmental risks. There must be a proper care for better understanding the site and preparation selection, well drilling, well production and leaving and restoration of the site. Each of these stages represents a potential contamination risk to environment. Almost two decades after the application of fracking, government gas institutions in the United States of America recognize and summarize the potential risks as the following [12, 27, 30, 34]:

4.2 Pressure in water supplies

The consumption of fresh water for drilling varies according to the characteristics of geological formation. Hydraulic fracturing process for a well can consume water in amounts from 9000 to 29,000 m^3 in a few days; such volume water demand could put water supply sources in the exploitation areas under pressure-affecting communities, economic activities and ecosystems [26].

Contamination in surface water and groundwater can occur, because groundwater quality can be affected if there are errors in wells construction that allow either infiltration of chemical additive injected from the surface, or leakage of return fluids coming out from the well. Surface water can be polluted by wastewater discharged before treatment, or by spills on soil in the storage and/or transport of return fluids or produced water. A deep storage method of wastewater disposal can contaminate both surface water and groundwater by spills streams, containing fractions of chemical additives and natural rock components (hydrocarbon, brine, heavy metals and radioactive elements). Hydraulic fracturing produces large volumes of wastewater at the surface, which may contain dissolved chemicals and other contaminants that require treatment before disposal or reuse.

4.3 Air pollution

There is emission of volatile organic compounds and toxic pollutants to air, by combustion and evaporative processes from the storage areas of reflux streams.

To analyze air pollution by toxic agents, studies report a significant concentration of benzene-toluene-ethylbenzene-xylene (BTEX), methylpyridine, butadiene, formaldehyde, among others, many of which are considered carcinogenic and neurotoxic and are emitted by the evaporative storage process or during the return of drilling fluids [35–38]. Air pollution also involves the use of motor trucks required to transport water, chemical additives, sand and materials; the number of trucks per well can be 320 to 1365 [28] or in some cases 1000 to 2000 [39]. There is release of natural gas during and after well drilling. The United States Environmental Protection Agency recognizes the problem and published a final rule in March 2024 that sets guidelines on the release of natural gas that will help reduce methane and other hazardous air pollutants in the oil and natural gas industry [40]. In addition, criteria pollutants CO, NO_2, SO_2 and $PM_{2.5}$ are reported in important concentrations, highlighting the concentrations of O_3 higher than the standard [28].

4.4 Ecosystem disturbance

Studies on potential environmental problems stating that [28, 35, 41–47] ecosystem disturbance put a life risk the original flora and fauna, because construction of platforms, where the drilling is carried out, requires an approximate area of 2 ha transforming the land use. The volumes used for rock fracturing are lost from the water cycle because of impact on water availability, recovering only 10–50% percentage of injected fluid, depending on the geological formation [29]. A possible aquifer contamination is related to the frequent injection of fluid volumes to the increasing number of wells making significant deformations in the shale, and those fluids can moved a mile or more to higher strata, unpredictably transporting pollutants through faults and fissures intersecting freshwater aquifers.

The returning flow by drilling contains chemical additives such as scale inhibitors, corrosion inhibitors, friction reducers, biocides, acids and lubricants. According to analysis of 944 commercial products, 632 substances have been identified; 353 of them have a potential effect on human health and because of this 75% affect skin exposure, eyes and other sensory organs as well as respiratory systems. About 40–50% of them affect the brain, kidney, and nervous, immune and cardiovascular systems; 37% affect the endocrine system and 25% can cause cancer and mutations [36]. At present, the exact composition of chemical additives is still unavailable information, because it is considered a trade secret, which makes unable to evaluate possible harm to humans and the environment, if they come into contact due to accidental causes or planned disposal [29, 48]. According to the United States Geological Survey (USGS), hydraulic fracturing can lead to small, non-significant earthquakes. However, the method of wastewater removal by injection in deep soil could produce larger and more damaging earthquakes.

5. Radiological aspects

5.1 Naturally Occurring Radioactive Materials (NORM)

Natural radionuclides from uranium and thorium series are present in the environment in soil, rocks, water and even food. ^{238}U and ^{232}Th are progenitors of those natural radioactive series, ending both in the stable element ^{206}Pb. The material containing these radioactive elements is called Naturally Occurring Radioactive Material

(NORM), which is present in mineral extraction industries including the oil industry whether conventional or unconventional [16, 49, 50]. The United States of America names NORM as them Technologically Enhanced, Naturally-Occurring Radioactive Material (TENORM), which clearly refers to natural radionuclides enriched by industrial activities [16]. **Figures 2** and **3** display natural radioisotopes from the decay chains ^{238}U and ^{232}Th and their mobility by transport (leaching and gas) during exploitation of gas and oil resources.

In oil and gas extraction processes, the water trapped in geological formations contains dissolved natural radionuclides, including ^{226}Ra, ^{228}Ra and ^{224}Ra together with their decay progeny in the reservoir rock. When oil and gas are extracted, these natural radionuclides present in this water reach the surface and due to changes in temperature and pressure, some precipitates or scales such as sulfate and calcium carbonate on the internal walls of the use equipment (pipes, valves, pumps, etc.) can accumulate significant concentrations of decay products ^{210}Pb and ^{228}Th [16, 51]. The activity concentrations in the scales present a wide range for ^{226}Ra, from less than 1000 Bq kg^{-1} up to 1,000,000 Bq kg^{-1} [50, 51]. These radionuclides can also appear in drilling mud streams and produced water, including equipment in the whole processes such as separators and storage tanks.

5.2 Environmental radioactivity by NORM

Pipes and equipment is therefore a source of radiological material [51], and workers can be subject to external gamma radiation exposure, or in particular cases to internal radiation by inhalation, ingestion or absorption of these radioactive material. Exposure to NORM does not cause severe effects similar to those associated with high

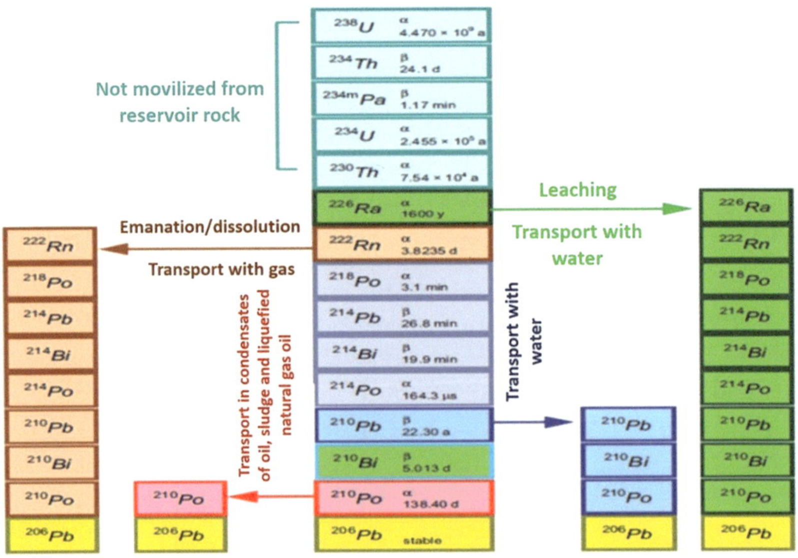

Figure 2.
^{238}U decay series (central column) and its mobility in the oil and gas industry. Leaching properties of ^{226}Ra permit its transport by water together with its decay products (right column). ^{222}Rn is transported by air and water due to its gaseous characteristic (left column). Modified from International Atomic Energy Agency. Radiation Protection and the Management of Radioactive Waste in the oil and gas industry. 2003.

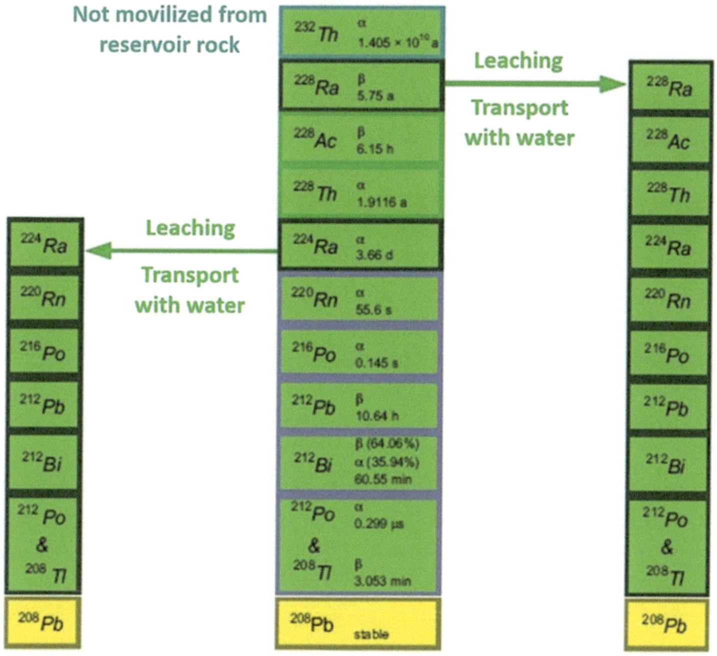

Figure 3.
^{232}Th decay series (central column) and its mobility in the oil and gas industry. Leaching properties of ^{228}Ra permit its transport by water together with its decay products (right column). ^{220}Rn is transported by air and water due to its gaseous characteristic (left column). Modified from International Atomic Energy Agency. Radiation Protection and the Management of Radioactive Waste in the oil and gas industry. 2003.

levels of radiation from man-made sources, but chronic exposure to NORM does; delayed effects of certain forms of cancer to general public can be present without following the adoption of appropriate safety practices.

The transport of natural radionuclides (gases or solids), during the exploitation of unconventional resources, comes from the forming rock, and the concentration and type of radionuclide depends mainly on the salinity of the water (connate water), associated with the hydrocarbon, the more saline the more radioactive, which is the characteristic of each geological formation [28, 44, 52–54]. Produced water treatments generate sludge that can concentrate natural radionuclides, so a specific treatment for NORM is required, because the concentration of radionuclides in the sludge of treatment plants is in many cases a limitation for the acceptance of treated waste disposed or reused in landfills on roads.

There are reports showing abnormal natural radionuclides associated with shale hydrocarbons. For the “Marcellus play” case in the United States of America, ^{226}Ra exceeds up to 267 times the safety limit in water for human consumption and ^{222}Rn in air is 70 times more than the natural background level [55]. Although ^{137}Cs in soil is not natural radionuclide, its concentration was 74 Bq kg^{-1}, because in “Marcellus play” case, ^{137}Cs was used as a tracer in the prospecting stage of shale gas, associated with contamination by rock cuttings scattered on the surface [56]. In Sweden, uranium content in waste exceeds 80 ppm needed a special permission for its management; radioactivity levels from Ra and Sr emissions in the fracturing fluids and cuttings have been confirmed and documented [43]. Estonia identified an increase in natural radionuclides during the power generation by plants using fuels from unconventional resources; in this case ^{210}Pb, ^{238}U, ^{226}Ra and ^{40}K were present in particles suspended in air and ash waste, from the process, in concentrations of 75.4, 75.8, 90.7 and 1201 Bq kg^{-1}, respectively [57, 58].

The International Atomic Energy Agency (IAEA) addresses the issue of NORM in order to harmonize the best practices in the industrial sector. Although NORM origin is known, the issue is recent and there are still tasks requiring greater precision such as management and the convenience of establishing centralized and specific facilities for the treatment and disposal of NORM waste. Part of the difficulty of management and control lies in the absence of NORM inventories that require the quantities of waste generated, concentrations of radioactivity, exposure rates and other hazardous materials of interest. Likewise, it is still necessary to improve the definition of methods and approaches for the sampling and characterization of NORM in industrial operations and in environment; there are another more specific areas such as local and beyond borders transport of NORM, remediation of contaminated sites and reuse options [59].

IAEA recognizes as a challenge the volumes of waste generated by unconventional wells in the oil and gas industry, which far exceeds those from conventional oil and gas wells. Therefore, there is a larger approaching for NORM waste revision, which depends not only on the origin of the oil and gas, but also on other factors such as geological formation, well age and salinity [50, 59].

However, in general, NORMs are controllable and do not represent a radiological emergency resulting in immediate danger to life. Protection to workers and public should consider long-term external exposure, ingestion of radioactive material, and inhalation of radon or thoron, descendants from radium [50].

6. Environmental radiological baseline: Mexico case

This section describes the importance of establishing an environmental baseline before any industrial activities, capable of NORM generation. This baseline will allow, after industrial activities, in determining if natural radionuclide concentrations are enhanced or not above the original background levels, and if so, whether they represent a possible radiological risk to the population.

6.1 Oil production and NORM

The United States of Mexico (Mexico) is located in the southern part of North America, bordered to the north by the United States of America, to the southeast by Belize and Guatemala, to the east by the Gulf of Mexico and the Caribbean Sea, and to the west by the Pacific Ocean [60]. The mineral extraction and processing industries including the oil industry are NORM generators; their distribution in the country is shown in the **Figure 4**.

Mexico started as an oil producer country in 1898; initially, the industry was running by foreign private companies; in 1938, the government took total control of the industry through the creation of the state company Petróleos Mexicanos. At the end of 2013, the laws were subject to an energy reform to allow private investments to participate in this energy sector, but maintaining the state company a high percentage of exploitation assignments [61]. Some justifications for that energy reform were to reverse the decrease in oil production from 3.5 X 10^6 barrels per day (bpd) in 2004 to 2.5 X 10^6 bpd in 2013. Addition consideration of that reform was reducing

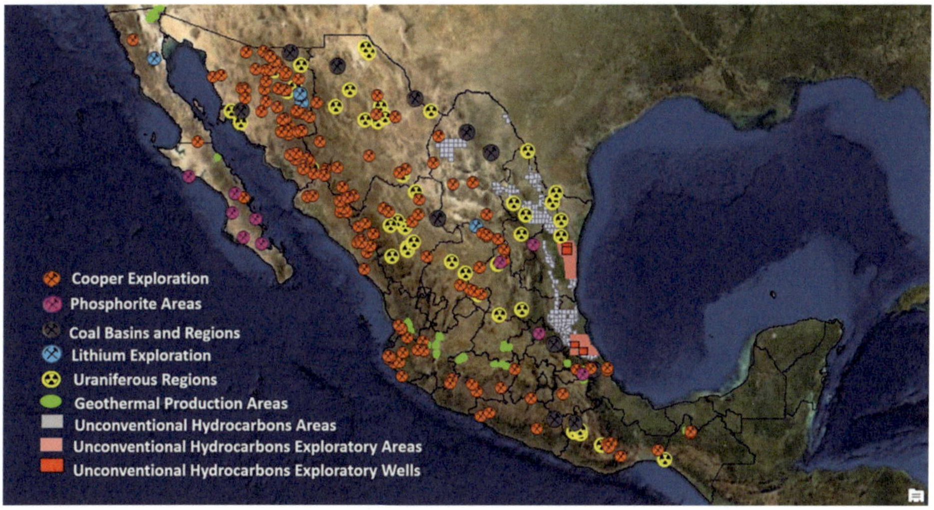

Figure 4.
Distribution of NORM-generating industry in Mexico. Mining and oil industries are present in the whole country.

gas imports, which could be achieved by oil and gas extraction from technical and financial capacity of private companies, including the unconventional oil resource extraction from shale rock, through fracking. The present government prohibited the fracking possibility in June 2019.

Geological provinces in Mexico cover areas of hundreds to thousands square kilometers, characterized by their lithology, structure and a series of geological evolution events. Geological provinces differ from each other by having stratigraphic or tectonic boundaries or both. In Mexico, there are fifty geological provinces, twelve have favorable conditions for hydrocarbons resources (oil provinces), and in them, six shales areas have been identified for potential exploration of unconventional gas reserves [11], see **Figure 5**.

It is estimated that Mexico has 64.2 billion barrels of oil equivalent (BBOE) of prospective unconventional shale resources. The relative amount to conventional resources is displayed in **Table 2** according to National Commission Hydrocarbons (Comisión Nacional de Hidrocarburos) [11]. This information results from 27 exploratory unconventional drilled wells, obtaining a success rate of around 85%.

The oil and mining industries are NORM generation activities in Mexico, which have not been so far subject to NORM regulation. Therefore, it was pertinent to assess the baseline due to natural radionuclide concentrations in four sites before drilling fracking exploratory wells. If fracking takes place in the future, there will be a comparison background for evaluating their long-term effect [35].

6.2 Study site

This first study on dose-rate baseline in Mexico refers to the radiation emission from natural radiactive chains ^{238}U and ^{232}Th [62]. The importance of each radionuclide in the environment depends on the concentration of activity, radiotoxicity and half-life.

This study took place in three zones, named Zone1, Zone2 and Zone3, at Tampico-Misantla oil province where 50.9% of unconventional resources are situated. The study

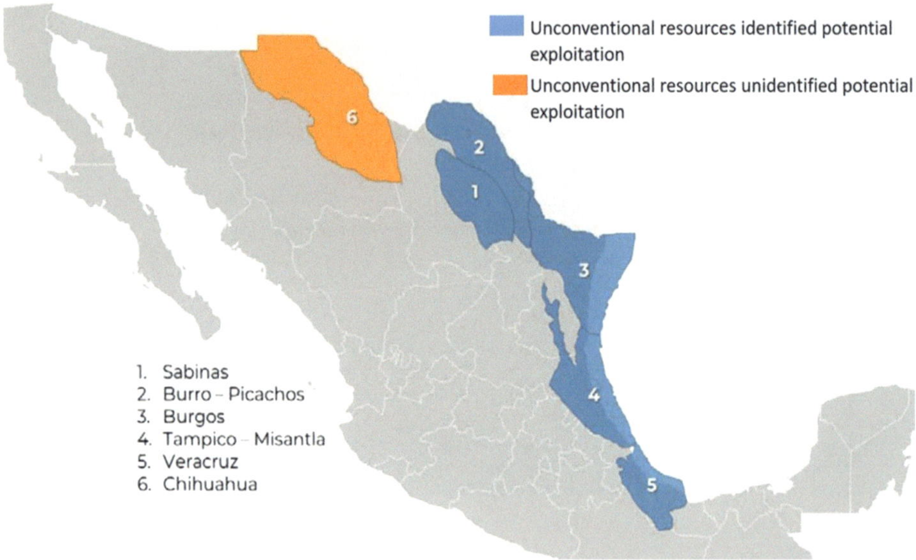

Figure 5.
Six provinces having unconventional hydrocarbon reserves. Modified from CNH, 2022 [11].

	Oil equivalent (BBOE)	Oil (Bb)	Natural gas (Tcf)
Unconventional	64.2	35.9	141.5
Conventional	48.7	32.1	83.2
Total	112.9	68.0	224.7

Table 2.
Relative conventional and unconventional hydrocarbon reserves in Mexico. Modified from CNH, 2022 [11].

in each of the three zones positioned a potential exploratory fracking well in the center of 1 km^2 area; 25 natural sampling points were located in a 200 x 200 m grid. The study considered three types of measurements: (a) gamma equivalent dose rate, (b) measurement of natural radionuclide concentrations in surface soil and (c) measurements of natural radionuclides in groundwater.

6.3 Equivalent dose rate

For gamma equivalent dose rate, measurements were taken at 1.5 m height, to associate them to human exposure from gamma-emitting radionuclides naturally present in the soil. *In situ* measurements were performed using a portable Ge(Hp) detector type ORTEC Microdetective HX, with a detection range of (0.05–10,000) $\mu Sv\ h^{-1}$ [62]. All measurements were geospatially interpolated using the krigging method of ArcGIS V. 10.7.1 software, which permitted to draw a dose rate radiological map for each three zones. The map was overlapped with layers containing population density, crop type and land use, with the aim for better identifying anomalous areas.

The maximum dose rate for all four sites was 1.83 $mSv\ a^{-1}$ and the minimum of 0.04 $mSv\ a^{-1}$; the dose rate accepted as environmental background by the United Nations Scientific Committee on the Effects of Atomic Radiation (UNSCEAR) [63]

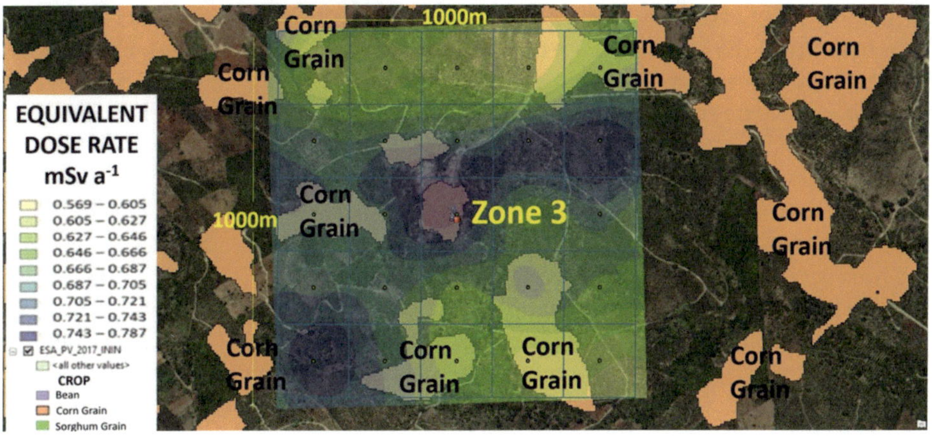

Figure 6.
Dose rate distribution in Zone3. Crop distribution (beans, corn and sorghum).

is in the range of 1–13 mSv a^{-1}. After this dose rate determination, three wells were drilled for fracking at an exploratory level; one of the wells inside the study site had already a history of conventional oil and gas exploitation (**Figure 6**).

6.4 Natural radionuclei in soil

Soil radionuclide measurements were from samples taken at a 0.20 m depth. Twenty-five samples were evaluated for each of the three zones. Each sample containing 1.5 kg of soil was sealed and labeled in polyethylene bags; in the laboratory samples were sieved to 100μm grain size and dried at 100°C for 24 h. Each sample was introduced into a 0.5-L Marinelli container for gamma spectrometry analysis using a Ge(Hp) detector with 30% efficiency [62]. This analysis allow identification and quantification of natural radionuclides; the results are shown in **Figure 7**.

^{226}Ra from the ^{238}U natural decay series and ^{228}Ra from the ^{232}Th natural decay series are the two most important radionuclides according to their mobility by water leaching properties, shown in **Figures 2** and **3**, respectively. For the three zones,

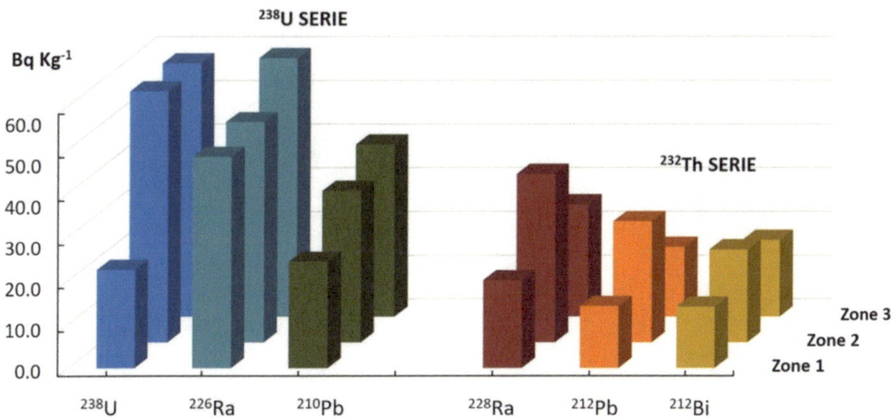

Figure 7.
Average activity concentrations of natural radionuclides in soil samples from the three zones.

the maximum concentration values are 59.2 ± 3.6 Bq kg^{-1} for ^{226}Ra and 38.5 ± 5.0 for ^{228}Ra, both are in the interval of average global concentration reported by UNSCEAR [63]. Activity concentrations of ^{238}U, ^{226}Ra and ^{210}Pb follow the same pattern for the three zones, which reveal the absence of industrial activity, and similar behavior is observed for ^{232}Th series.

6.5 Presence of radioactivity in water

Three parameters were evaluated in groundwater samples: (a) gross alpha, (b) gross beta and (c) activity concentrations of natural radionuclide ^{238}U and ^{226}Ra. The water samples were collected from aquifers (Zone1 and Zone2). The aquifers were identified regarding the average annual water availability, direction of its flow and their use (livestock, agriculture, domestic). Groundwater samples were obtained from 35 domestic wells, see **Figures 8** and **9**.

Each water sample contained 5 L volume, electrical conductivity was recorded and samples were acidified to a pH 2. In the laboratory, water samples were reduced by evaporation to 0.45 L in volume, introduced into a 0.5-L Marinelli container for gamma spectrometry analysis using a Ge(Hp) detector with 30% efficiency, to determine the radionuclides using gamma spectrometry [62]. Another set of samples of 0.45 L in volume was even more reduced to 10 mL for assessing gross alpha and gross beta, using a proportional flow detector PROTEAN model WPC-1050 [62]; **Table 3** display results.

^{238}U and ^{226}Ra, concentrations in groundwater did not exceed the minimum detectable limit and total alpha and beta emissions in groundwater samples from these sites are all below 0.5 and 1.0 Bq L^{-1}, respectively, which corresponds to the

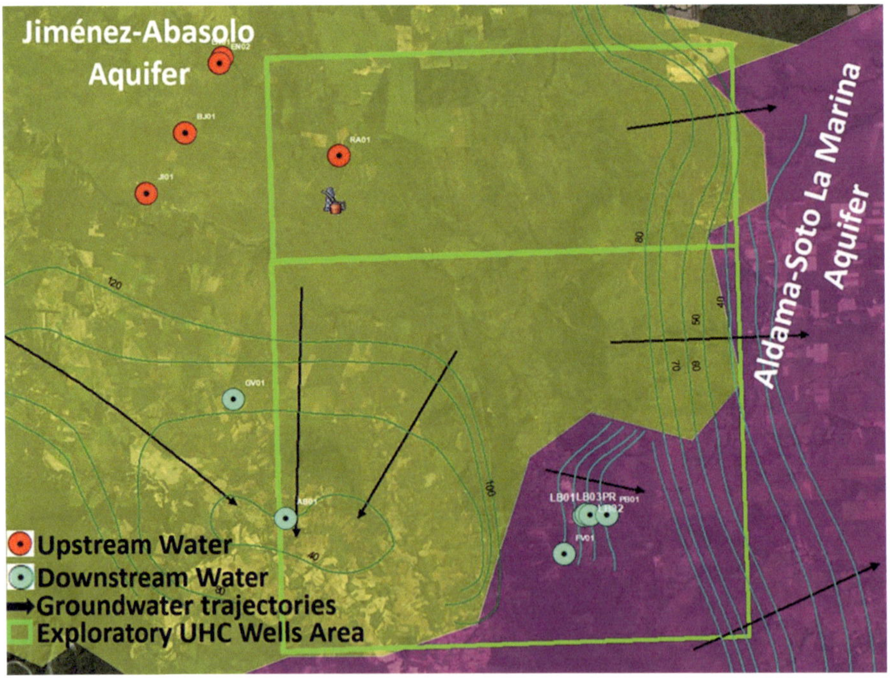

Figure 8.
Water sampling from 10 water wells at Zone1. Two aquifers and water trajectories are identified.

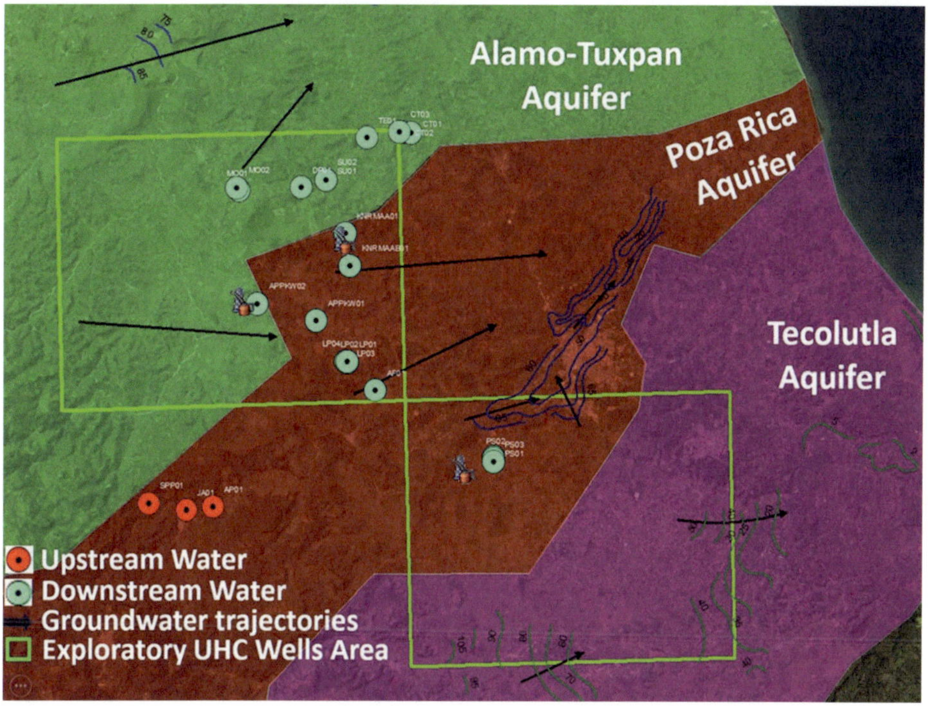

Figure 9.
Water sampling from 20 water wells at Zone2. Three aquifers and water trajectories are identified.

Parameter	Zone1 (Bq L^{-1})	Zone2 (Bq L^{-1})	Reference level* (Bq L^{-1})
Gross alpha	ND – 0.07	ND	0.5
Gross beta	ND – 0.48	ND – 0.2	0.5
Activity concentration ^{238}U	≤1.3	≤1.3	—
Activity concentration ^{226}Ra	≤1.0	≤1.0	—
Groundwater samples	15	20	—

**World Health Organization-guideline values drinking water [62], ND: non-detected.*

Table 3.
Activity concentrations in groundwater samples.

maximum suggested limits established by the World Health Organization (WHO) for drinking water.

6.6 Water stress

In terms of environmental radiological matters, unexploited areas with shale reserves do not currently present any radiological risk to the environment and human beings. If fracking takes place, it is important to define how much time later will be necessary to make a comparison, since there are studies performed a year later that do not reflect a significant variation [35, 64].

The interest in finding a baseline for water is due to the fragile conditions of water resource. The World Resources Institute (WRI), in a study carried out in

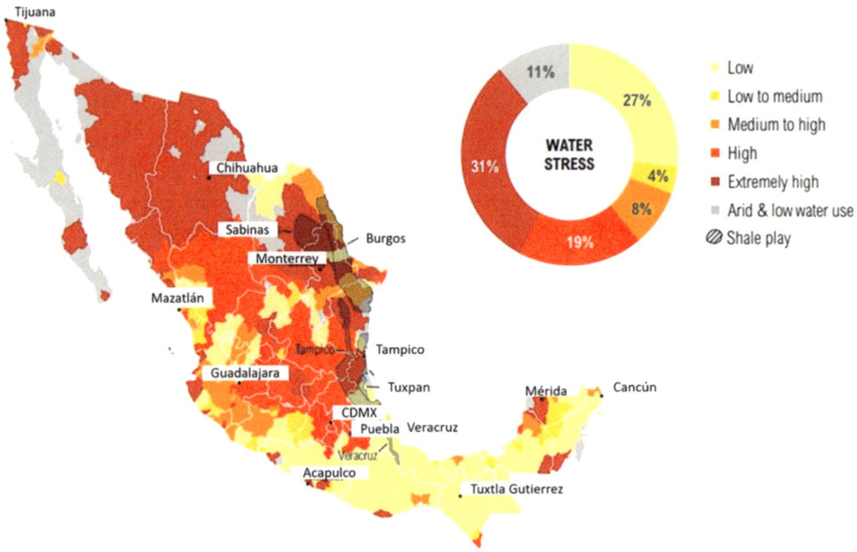

Figure 10.
Regions of Mexico with shale reserves and water stress level. Adapted from World Resources Institute (WRI). Global Shale Gas Development: Water Availability and Business Risks. 2014.

2014, presents the baseline of water stress in areas where there is potential exploitation of unconventional resources. Water stress is the relationship between the total use of water by municipal, industrial and agricultural users and the available renewable surface water. High-water stress values can indicate greater competition among users and greater depletion of water resources [65]. According to this study, 31% of shale resources are in regions with high to extremely high-water stress see **Table 2** for countries with the largest reserves of this resource and **Figure 10** for Mexico.

7. Conclusions

The expected low-level radioactive results define the baseline for the three zones in Mexico, before fracking activities. There is no radiological risk due to equivalent dose rate; the maximum dose rate was 1.83 mSv a^{-1}, which is in the acceptable background interval (1–13) mSv a^{-1} [63]. ^{226}Ra and ^{228}Ra the most mobil radionuclei in soil are both in the average global concentration reported by UNCEAR [63]. Measurements of gross alpha and gross beta in water did not exceed the maximum suggested limits by the World Health Organization [62] for drinking water. Water stress in Mexico is an important issue before considering fracking activities.

The radiological aspects reported for the Marcellus formation in the United States of America [15] motivate the first baseline measurement in Mexico due to natural radionuclides in soil and water for areas not just before fracking, but areas where extractive industries potentially could enhance Naturally Occurring Radiactive Materials.

Attention should be on the International Atomic Energy Agency [60] suggestion that puts forward the regulation needs for the enhanced of Naturally Occurring Radiactive Materials in wastewater and sludge derived from fracking technique.

Large volumes of water are needed, which need a careful analysis of the fracking technique, especially in those areas where high-water stress is an environmental issue, as pointed out by the World Resources Institute [66].

Acknowledgements

We thank to International Atomic Energy Agency through MEX 7013 project for technical assistance to perform baseline criteria, to Environmental Radiological Laboratory at Instituto Nacional de Investigaciones Nucleares for technical support and to National Oil Company PEMEX for permission to perform the radiological baseline.

Author details

Beatriz Hernandez-Mendez[1], Sergio Suarez-Contreras[1], Juan Carlos Sanchez-Meza[2], Edith E. Gutierrez-Segura[2] and Miguel Balcazar[1*]

1 Instituto Nacional de Investigaciones Nucleares, Ocoyoacac Estado de México, México

2 Universidad Autónoma del Estado de México, Toluca, Estado de México, México

*Address all correspondence to: miguel.balcazar@inin.gob.mx

IntechOpen

© 2025 The Author(s). Licensee IntechOpen. This chapter is distributed under the terms of the Creative Commons Attribution License (http://creativecommons.org/licenses/by/4.0), which permits unrestricted use, distribution, and reproduction in any medium, provided the original work is properly cited.

References

[1] International Energy Agency (IEA). World Energy Outlook 2023. Paris France: IEA Publications; 2023. 335 p

[2] UNAM Global Revista. Transición energética para una sociedad justa y sostenible [Internet]. 2023. Available from: https://unamglobal.unam.mx/global_revista/transicion-energética-para-una-sociedad-justa-y-sostenible/ [Accessed: July 02, 2024]

[3] United Nations Climate Change. El Acuerdo de París. [Internet] 2024. Available from: https://unfccc.int/es/acerca-de-las-ndc/el-acuerdo-de-paris [Accessed: July 02, 2024]

[4] Rowlatt J, BBC News. Combustibles fósiles: ¿puede realmente la humanidad terminar con su dependencia de ellos? [Internet]. 2023. Available from: https://www.bbc.com/mundo/articles/cevqlyygw0qo [Accessed: July 08, 2024]

[5] United Nations Environment Programme. An Eye on Methane: International Methane Emissions Observatory 2022 Report. Available from: https://wedocs.unep.org/20.500.11822/40864 [Accessed: July 08, 2024]

[6] Caineng Z, Guosheng Z, Zhi Y, Shizhen T, Lianhua H, Rukai Z, et al. Concepts, characteristics, potential and technology of unconventional hydrocarbons: On unconventional petroleum geology. Petroleum Exploration and Development. 2013;**40**(4):413-428

[7] U.S. Energy Information Administration (EIA). Short-Term Energy Outlook. Washington, DC. USA; 2024. 56 p. Available from: https://www.eia.gov/outlooks/steo/

[8] U.S. Energy Information Administration. How much shale gas is produced in the United States? [Internet]. 2024. Available from: https://www.eia.gov/tools/faqs/faq.php?id=907&t=8#:~:text=In%202023%2C%20about%2077%25%20(,production%20was%20from%20shale%20formations.&text=Last%20updated%3A%20April%2029%2C%2020-24%3B%20preliminary%20data [Accessed: August 06, 2024]

[9] U.S. Energy Information Administration (EIA). Annual Energy Outlook. AEO2023 Narrative. Washington, DC, USA; 2023. 50 p

[10] U. S. Energy Information Administration. World Shale Resource Assessments [Internet]. Last updated: September 24 2015. Available from: https://www.eia.gov/analysis/studies/worldshalegas/ [Accessed: July 09, 2024]

[11] Comisión Nacional de Hidrocarburos (CNH). Retos y Oportunidades de la Producción de Petróleo y Gas Natural de Yacimientos No Convencionales en México. Ciudad de México, México: CNH; 2022. 84 p

[12] Hydraulic Fracturing & Health [Internet] 2022. Available from: https://www.niehs.nih.gov/health/topics/agents/fracking [Accessed: July 09, 2024]

[13] Li L, Dominici F, Blomberg AJ, Bargagli-Stoffi FJ, Schwartz JD, Coull BA, et al. Exposure to unconventional oil and gas development and all-cause mortality in medicare beneficiaries. Nature Energy. 2022 Feb;7(2):177-185. DOI: 10.1038/s41560-021-00970-y Epub 2022 Jan 27

[14] Howarth RW. Ideas and perspectives: Is shale gas a major driver of recent

increase in global atmospheric methane? Biogeosciences. 2019;**16**:3033-3046. DOI: 10.5194/bg-16-3033-2019

[15] Department of Environmental Protection of Pennsylvania. PA (DEPP), Bureau of Air Quality. Northeastern Pennsylvania Marcellus Shale Short-Term Ambient Air Sampling Report. Department of Environmental Protection of Pennsylvania; 2011. 48 p

[16] International Atomic Energy Agency (IAEA). Extent of Environmental Contamination by Naturally Occurring Radioactive Material (NORM) and Technological Options for Mitigation. Technical Report Series No. 419. Vienna: International Atomic Energy Agency; 2003. 208 p

[17] Instituto Mexicano del Petróleo (IMP). Biblioteca Visual del Petróleo. 2nd ed. México, DF: IMP; 2015. 13 p

[18] Origen del petróleo e historia de la perforación en México. [Internet]. 2024. Available from: file:///D:/Origen%20del%20petroleo%20e%20historia.pdf [Accessed: August 13, 2024]

[19] Energy glossary [Internet]. 2024. Available from: https://glossary.slb.com/ [Accessed: August 13, 2024]

[20] Soeder DJ, Borglum SJ. The Fossil Fuel Revolution: Shale Gas and Tight Oil. South Dakota, SD, United States: Elsevier; 2019. 336 p. DOI: 10.1016/C2017-0-03254-8

[21] Soeder DJ. The successful development of gas and oil resources from shales in North America. Journal of Petroleum Science and Engineering. 2018;**163**:399-420. DOI: 10.1016/j.petrol.2017.12.084

[22] Zee Ma Y, Holditch SA. Unconventional Oil and Gas Resources Handbook: Evaluation and Development. Oxford, UK: Gulf Professional Publishing Elsevier; 2015. 536 p. ISBN-10: 9780128022382

[23] Comisión Nacional de Hidrocarburos (CNH). El Sector del Gas Natural: Algunas Propuestas para el Desarrollo de la Industria Nacional. Ciudad de México, México: CNH; 2018. 189 p

[24] What is "Unconventional" Oil and Gas?. [Internet]. 2024. Available from: https://www.planete-energies.com/en/media/article/what-unconventional-oil-and-gas [Accessed: August 06, 2024]

[25] Fractura Hidraúlica. Fracking. Video Informativo sobre la técnica del Shale. IAPG Channel. [Internet]. 2017. Available from: https://www.youtube.com/watch?v=TVZ0DOZ59SA [Accessed: July 12, 2024]

[26] Esto es lo que debes saber sobre el fracking y sus riesgos [Internet]. 2022. Available from: https://www.wwf.org.mx/?378492/Esto-es-lo-que-debes-saber-sobre-el-fracking-y-sus-riesgos [Accessed: July 12, 2024]

[27] Natural Gas Explained. Natural gas and the environment [Internet]. 2024. Available from: https://www.eia.gov/energyexplained/natural-gas/natural-gas-and-the-environment.php [Accessed: August 13, 2024]

[28] United States Goverment Accountability Office (USGAO). Oil and Gas. Information on Shale Resources Development, and Environmental and Public Health Risks. Report to Congressional Requesters. GAO-12-732. Washington DC. 2012. 66 p

[29] Concerned Health Professionals of New York and Physicians for Social Responsibility. Compendium of Scientific, Medical, and Media

Findings Demonstrating Risks and Harms of Fracking and Associated Gas and Oil Infrastructure. 8th ed2022. 577 p. Available from: http://concernedhealthny.org/compendium/

[30] Hydraulic fracturing [Internet]. 2019. Available from: https://www.usgs.gov/mission-areas/water-resources/science/hydraulic-fracturing [Accessed: July 20, 2024]

[31] Carmalt SW, Moscariello A. Environmental issues related to fracking. The Journal of World Energy Law & Business. 2020;**13**(5-6):386-410. DOI: 10.1093/jwelb/jwaa039

[32] Michaels C, Simpson JL, Wegner W. Fractured Communities Case Studies of the Environmental Impacts of Industrial Gas Drilling. Ossining. NY, USA: Riverkeeper; 2010. 39 p

[33] Auer PS. Understanding Naturally Occurring Radioactive Material in the Marcellus Shale. In: A discussion of the naturally occurring radioactive material resulting from natural gas drilling in the Marcellus Shale, The Science Beneath the Suarface. Issue Number 4. New York: Paleontological Research Institution. Marcellus Shale; 2011. 8 p

[34] The process of unconventional natural gas production [Internet]. 2024. Available from: https://www.epa.gov/uog/process-unconventional-natural-gas-production [Accessed: August 14, 2024]

[35] Werner AK, Vink S, Watt K, Jagals P. Environmental health impacts of unconventional natural gas development: A review of the current strength of evidence. Science of the Total Environment. 2015;**505**:1127-1141

[36] Wolf Eagle Environmental. Town of Dish, Texas ambient air monitoring analysis final report. 2009. 4 p

[37] Colborn T, Schultz K, Herrick L, Kwiatkowski C. An exploratory study of air quality near natural gas operations. TEDX, the endocrine disruption exchange. Human and ecological risk assessment. An Intenational Journal. 2012;**17**(5):1039-1056

[38] Meng Q. Spatial analysis of environment and population at risk of natural gas fracking in the state of Pennsylvania, USA. Science of the Total Environment. 2015;**515-516**:198-206

[39] McDermott-Levy R, Kaktins N. Preserving health in the Marcellus region. The Pennsylvania Nurse. 2012;**67**:4-12

[40] Environmental Protection Agency (EPA). Standards of performance for new, reconstructed, and modified sources and emissions guidelines for existing sources: Oil and natural gas sector climate review. Final Rule. 40 CFR Part 60 [EPA–HQ–OAR–2021-0317; FRL–8510-01– OAR] RIN 2060–AV16. Federal Register Rules and Regulations. 2024;**89**(47). 408 p

[41] Broomfield M. Support to the identification of potential risk for the environment and human health arising from hydrocarbons operations involving hydraulic fracturing in Europe. In: Report for European Comission DG Environment. Oxfordshire, UK: AEA Tecnology Plc. AEA/R/ED57281; Issue Number 17, 2012. 292 p

[42] Lechtenböhmer S, Altmann M, Capito S, Matra Z, Weindrorf W, Zittel W. Impacts of shale gas and shale oil extraction on the environment and on human health. Committee on Environment Public Health and Food Safety. 2011. Available from: http://www.europarl.europa.eu/document/activities/cont/201107/20110715ATT24183/20110715ATT24183EN.pdf [Accessed: July 02, 2024]

[43] Sovacool BK. Cornucopia or curse? Reviewing the costs and benefits of shale gas hydraulic fracturing (fracking). Renewable and Sustainable Energy Reviews. 2014;**37**:249-264

[44] Healy D. Hydraulic fracturing or "Fracking": A short summary of current knowledge and potential environmental impacts. In: Report Environmental Protection Agency Ireland. UK: Science Technology, Research & Innovation for the Environments; 2012. 26 p

[45] DNV. Risk management of shale gas developments and operations. 2013. DNV-RP-U301

[46] International Energy Agency (IEA). Golden rules for a golden age of gas. 2012. Available from: http://www.worldenergyoutlook.org/media/weowebsite/2012/goldenrules/Golden_Rules_ES_Spanish_WEB.pdf

[47] American Petroleum Institute (API). Estimate of Impacts of EPA Proposals to Reduce Air Emissions from Hydraulic Fracturing Operations. Advanced Resources International Inc; 2012. Available from: https://obamawhitehouse.archives.gov/sites/default/files/omb/assets/oira_2060/2060_03222012-4.pdf [Accessed: August 08, 2024]

[48] Speight JG. Deep Shale Oil and Gas. Oxford, United Kingdom: Gulf Professional Publishing. Elsevier; 2016. 492 p

[49] Dwipayana CAW, Moersidik SS, Pratama MA. Estimation radiation dose from operation of petroleum NORM waste disposal in landfill using TSD-DOSE. The ninth international conference on theoretical and applied physics (ICTAP). Journal of Physics: Conference Series. 2020;**1572**:012031. DOI: 10.1088/1742-6596/1572/1/012031

[50] International Commission on Radiological Protection (ICRP). Radiological protection from naturally occurring radioactive material (NORM) in industrial processes. ICRP Publication 142; 2019. Ann. ICRP 48(4)

[51] International Atomic Energy Agency (IAEA). Radiation Protection and the Management of Radioactive Waste in the Oil and Gas Industry. Vienna: AIEA Safety Report Series. No. 34; 2003. 130p

[52] Fair JH. Radionuclides in fracking wastewater. Managing a toxic blend. Environmental Health Perpesctives. 2014;**122**:A50-A55

[53] Witter R, Stinson K, Sackett H, Putter S, Kinney G, Teitelbaum D, et al. Potential Exposure Related Human Health Effects of Oil and Gas Development: A White Paper. United States of America. Denver: Colorado School of Public Health University of Colorado Denver; 2008. 16 p

[54] House P. Hydraulic Fracturing: Impacts on the Environment and Human Health. Northern Ireland: Chartered Institute of Environmental Health; 2012. 12 p

[55] Resnikoff M. Radon in natural gas from marcellus shale. Ethics in Biology, Engineering and Medicine. 2011;**2**(4):317-331. DOI: 10.1615/ethicsbiologyengm

[56] New York State Department of Environmental Conservation (DEC). Natural gas development activities & high-volume hydraulic fracturing. In: Preliminary Revised Draft Supplemental Generic Environmental Impact Statement On The Oil, Gas and Solution Mining Regulatory Program. Well Permit Issuance for Horizontal Drilling and High-Volume Hydraulic Fracturing to

Develop the Marcellus Shale and Other Low-Permeability Gas Reservoirs. Albany NY: NYS DEC Division of Mineral Resources; 2011. pp. 4-144

[57] Vaasma T et al. The enrichment of natural radionuclides in oil shale-fired power plants in Estonia - the impact of new circulating fluidized bed technology. Journal of Environmental Radioactivity. 2014;**129**:133-139

[58] Vaasma T, Kiisk M, Meriste T, Tkaczyk AH. The enrichment behavior of natural radionuclides in pulverized oil shale-fired power plants. Journal of Environmental Radioactivity. 2014;**138**:427-433. DOI: 10.1016/j.jenvrad.2014.02.027

[59] International Atomic Energy Agency (IAEA). Management of naturally occurring radioactive material (NORM) in industry. In: Proceedings of an International Conference Vienna, Austria, 18-30 October 2020. Proceedings Series. Austria; 2022. 80 p

[60] Información General Sobre México. [Internet]. 2013. Available from: https://embamex.sre.gob.mx/republicadominicana/index.php/avisos/2-uncategorised/127-informacion-general-sobre-mexico [Accessed: July 16, 2024]

[61] La industria petrolera en México, cuestión de tiempo [Internet]. 2018. Available from: https://www.anahuac.mx/generacion-anahuac/la-industria-petrolera-en-mexico-cuestion-de-tiempo [Accessed: July 16, 2024]

[62] Hernández-Méndez B, Balcázar M, Ángeles-Carranza A, Suárez-Contreras S, Quintero-Ponce E, Bárcenas-Robles M, et al. Environmental radiological baseline in unconventional oil and gas areas of Mexico. Radiation Protection Dosimetry. 2023;**199**(18):2189-2193. DOI: 10.1093/rpd/ncad257

[63] United Nations Scientific Committee on the Effects of Atomic Radiation (UNSCEAR). Sources and Effects of Ionizing Radiation. UNSCEAR 2008 Report to the General Assembly with Scientific Annexes. Vol. I. New York. USA: United Nations Publication; 2010

[64] Nelson AW, Knight AW, Eitrheim ES, Schultz MK. Monitoring radionuclides in subsurface drinking water sources near unconventional drilling operations: A pilot study. Journal of Environmental Radioactivity. 2015;**142**:24-28

[65] World Resources Institute (WRI). Global Shale Gas Development: Water Availability and Business Risks. Washington, DC, USA; 2014 88 p

[66] Reig P, Maddocks A. World Resources Institute (WRI) 40 percent of countries with largest shale energy resources face water sress. 2014. Available from: https://www.wri.org/insights/40-percent-countries-largest-shale-energy-resources-face-water-stress [Accessed: July 16, 2024]

Section 3

Infrastructure and Technology Innovations

Chapter 8

Perspective Chapter: GIS and Remote Sensing in Assessing Interdependencies within Oil and Gas Infrastructure

Rifaat Abdalla

Abstract

The chapter provides a conceptual model rather than a complete analysis case study-based approach to comprehensively assess interdependencies within the oil and gas sector. Delving into the intricate connections among pipelines, refineries, drilling operations, and transportation networks, the chapter elaborates on the utility of advanced GIS and remote sensing techniques rather than employing them directly in assessing interdependencies within the oil and gas sector. Emphasizing the significance of collaborative data sharing, the chapter visualizes relationships and elucidates the imperative of proactive management practices to fortify resilience. Through detailed case studies and practical instances, it vividly illustrates the possible tangible outcomes of such analysis, offering valuable insights into emergency preparedness, risk mitigation, and resource allocation optimization within the industry. Catering to professionals, researchers, and stakeholders navigating the complex web of dependencies within oil and gas infrastructure, this chapter serves as an essential resource. It ensures stability and uninterrupted functionality, even in the face of unforeseen emergencies and disruptions, by facilitating a profound understanding of the connections and aiding in the implementation of effective strategies to manage them effectively.

Keywords: infrastructure, interdependence, oil and gas, risk, visualization, modeling

1. Introduction

In the intricate tapestry of the oil and gas industry, the concept of Interdependency weaves an indispensable narrative, shaping the core of its operations and functionality. At the heart of this multifaceted sector lies an intricate network of components and systems, meticulously intertwined to form the backbone of oil and gas facilities. From sprawling pipelines traversing vast distances to intricate refineries processing raw resources, from the intricacies of drilling operations to the labyrinthine transportation networks spanning continents—each element represents a crucial thread in the larger fabric of the industry [1].

The interplay among these diverse components creates a web of dependencies, establishing vital connections and relationships that sustain the sector's operations. However, this Interdependency fosters efficiency and productivity but also introduces a precarious balance. In this delicate ecosystem, the failure or disruption of one element can reverberate across the entirety of the oil and gas infrastructure. Such interconnectedness, while offering unparalleled synergy, also poses a complex challenge, as any disturbance within this system could potentially cascade into profound consequences, impacting not only individual facilities but the stability and resilience of the industry at large.

This chapter embarks on a comprehensive exploration of these intricacies, delving into the fundamental significance of Interdependency within oil and gas facilities. By dissecting the various components, systems, and their interconnected nature, we aim to illuminate the intricate relationships that underpin the sector's operations. From understanding how pipelines rely on refineries for processing to the symbiotic relationship between drilling operations and transportation networks, our journey navigates the labyrinth of dependencies that shape the industry's landscape.

Moreover, this exploration is an essential foundation for comprehending the challenges and vulnerabilities inherent in this web of interdependencies. By acknowledging the potential vulnerabilities and consequences of systemic disruptions, we seek to identify areas of concern and pave the way for strategic measures and resilient frameworks that fortify the sector against unforeseen risks.

In this expedition, we unravel the interconnected tapestry of oil and gas facilities, illuminating the significance of Interdependency and its profound implications on the stability, efficiency, and resilience of this vital industry [2].

The chapter focuses on a comprehensive exploration of the intricate interconnections within the oil and gas infrastructure, specifically emphasizing interdependency analysis. This process involves scrutinizing and understanding the complex relationships and interconnections among various components of oil and gas facilities. These components may include pipelines, refineries, drilling operations, and transportation networks. The goal is to shed light on the fundamental concept of Interdependency and its contextual significance within the sector.

Given the nature of the research, it is assumed that the lack of specific data related to this field necessitates the development of topics using hypothetical scenarios. In other words, the chapter utilizes hypothetical situations and scenarios to illustrate and analyze the potential interdependencies within oil and gas facilities. These hypothetical scenarios serve as a means to explore and discuss the interwoven reliance among different infrastructure elements, considering various potential challenges and disruptions [3].

The interdependency analysis is crucial for understanding how the relationships among different constituents of oil and gas infrastructure profoundly impact the stability of the entire sector. By delving into the hypothetical scenarios, the chapter may highlight the complexities in quantifying, visualizing, and managing these intricate relationships. The discussion could also touch upon the challenges inherent in conducting interdependency analysis due to the unavailability of real-world data, emphasizing the importance of proactive management practices to fortify and elevate the resilience of these critical facilities. Overall, the chapter is anticipated to contribute insights into the nuanced evaluation and analysis of interdependencies within oil and gas infrastructure, offering valuable perspectives on how the sector can navigate challenges and disruptions [4].

2. Methods and tools for assessing interdependencies

Evaluating interdependencies within the intricate framework of oil and gas facilities demands a nuanced approach, utilizing various methods and tools to comprehend the complex relationships binding multiple infrastructure elements. We delve into the diverse arsenal of methodologies, encompassing network analysis, system modeling, and data-driven approaches, all aimed at unraveling the intricate web of dependencies within the industry [5].

2.1 Network analysis

Within the context of oil and gas facilities, network analysis is a powerful tool to decode the intricate relationships and dependencies among various infrastructure elements. It involves the construction of graphical models representing the interconnectedness of components like pipelines, refineries, drilling sites, and transportation networks. Each element within this network is depicted as a node, while the connections or dependencies between these elements are represented as edges or links. By employing mathematical algorithms and graph theory principles, network analysis offers a comprehensive visualization of these intricate interconnections, revealing the flow of resources, dependencies, and potential vulnerabilities within the system [6].

One of the key aspects of network analysis is its ability to identify critical nodes or central elements within the infrastructure network. Centrality measures such as integration and centrality or degree centrality highlight the nodes that serve as crucial hubs or pathways through which resource flows or information traverse. Understanding these critical points enables stakeholders to prioritize them for monitoring, maintenance, or fortification, enhancing the system's overall resilience [7].

Furthermore, vulnerability assessments through network analysis help identify elements that, if disrupted, might cause significant impacts on the overall functionality of the system. By simulating potential failures or disruptions and assessing their ripple effects through the network, stakeholders can proactively devise strategies to mitigate risks, strengthen redundancies, and develop contingency plans to minimize the potential impacts of such disruptions (**Figure 1**) [7, 8].

The visual representation of the network model aids in comprehending the complexities of the oil and gas infrastructure, allowing stakeholders to grasp the relationships and dependencies spatially and logically. This visual insight is crucial for decision-makers in formulating effective strategies for optimizing operations, improving safety, and ensuring the robustness of the entire oil and gas ecosystem.

Network Analysis is a pivotal tool in comprehending the relationships and dependencies among different components of oil and gas facilities. By employing graph theory and mathematical models, network analysis enables the visualization and quantification of interconnections among pipelines, refineries, drilling operations, and transportation networks. Mapping these connections provides invaluable insights into the critical nodes and vulnerabilities within the system, facilitating a deeper understanding of how disruptions in one area may reverberate throughout the entire network (**Figure 2**) [9, 10].

2.2 System modeling

System modeling is a robust methodology within interdependency analysis in oil and gas facilities, offering a sophisticated framework to comprehend the complex interactions and behaviors of the interconnected infrastructure elements. These

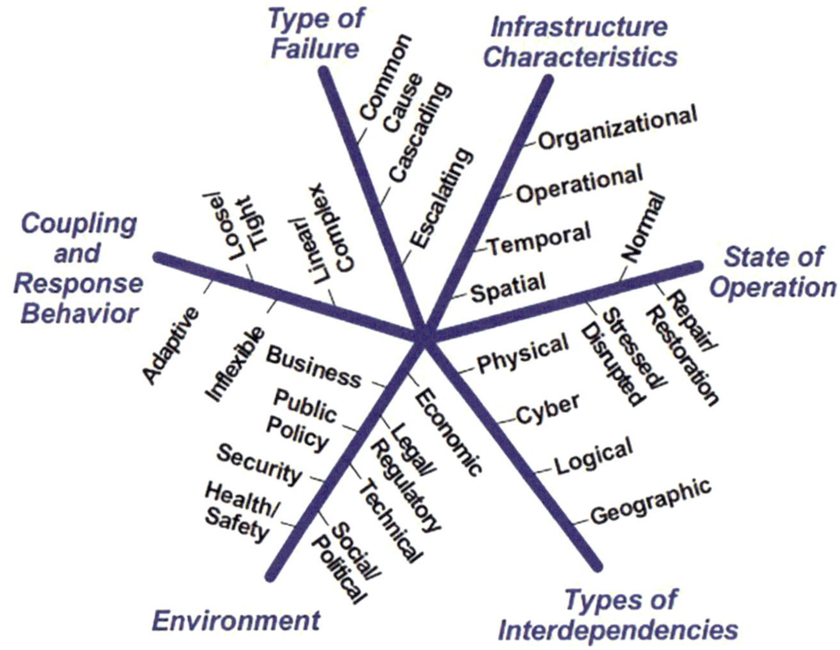

Figure 1.
Dimensions for describing infrastructure interdependency after [8].

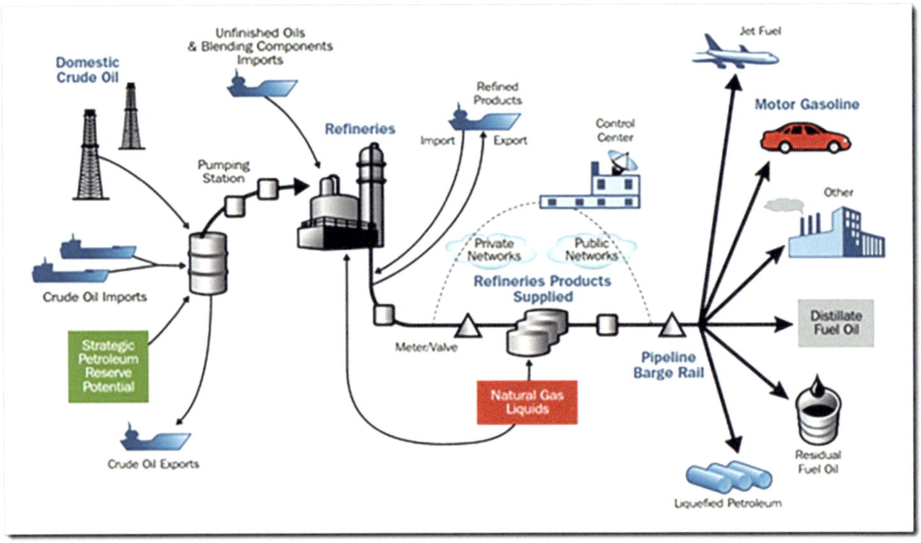

Figure 2.
The interdependence of oil infrastructure with other critical sectors after [9].

models, often built using advanced computational techniques and mathematical algorithms, aim to encapsulate the intricate relationships among pipelines, refineries, drilling operations, and transportation networks.

The primary function of system modeling lies in its ability to simulate the behaviors and dynamics of the entire oil and gas ecosystem. By integrating various data inputs, including operational parameters, geographical information, resource flows, and technological dependencies, these models create virtual representations that mirror the real-world functionalities of the infrastructure. This comprehensive simulation lets stakeholders understand how changes or disturbances in one system segment might propagate through the interconnected network [11].

2.3 Scenario analysis

A crucial aspect of system modeling enables stakeholders to explore "what-if" scenarios and assess their potential impacts on the entire system. Stakeholders can observe and evaluate the ripple effects in the whole oil and gas infrastructure by introducing simulated disruptions, changes in operational conditions, or failures in specific components. This analytical approach provides a deeper understanding of the interconnectedness and dependencies within the system, illuminating potential vulnerabilities and critical points susceptible to disruptions.

Furthermore, system modeling facilitates the identification of potential bottlenecks or areas of heightened risk within the infrastructure. Through sensitivity analysis and simulation of various scenarios, stakeholders can pinpoint critical nodes or segments where failures or disturbances could have significant cascading effects. Armed with this knowledge, decision-makers can proactively devise strategies to reinforce these vulnerable points, enhance redundancies, and develop contingency plans to minimize the impacts of potential disruptions.

System modeling offers a comprehensive and dynamic view of the interactions among diverse elements within oil and gas facilities. It empowers stakeholders to anticipate, understand, and mitigate the impacts of disruptions or changes within the system, enabling informed decision-making and fostering the resilience of the oil and gas industry in the face of evolving challenges.

2.4 Geospatial visualization of interdependencies

Geospatial visualization is a powerful lens to comprehend and analyze the intricate interdependencies within oil and gas facilities. Geospatial visualization provides a comprehensive and spatially contextualized understanding of the relationships between various infrastructure elements by overlaying geographical data onto complicated network analyzes and system models. This approach enables stakeholders to map out the physical locations of pipelines, refineries, drilling sites, and transportation networks, thereby visualizing their spatial relationships and dependencies. Through interactive maps and spatial representations, geospatial visualization offers a dynamic perspective, facilitating the identification of critical nodes, vulnerabilities, and potential impact zones across diverse terrains and regions. This spatial awareness enhances the precision of risk assessments and empowers decision-makers to devise targeted strategies for fortifying the resilience of oil and gas facilities in specific geographic contexts [2].

However, quantifying and visualizing these relationships present formidable challenges. The complexities inherent in the industry, such as the dynamic nature of operations, diverse technologies, and the sheer scale of infrastructure, pose hurdles in accurately assessing interdependencies. Variables such as geographical dispersion, various operational protocols, and evolving technologies further complicate the creation of comprehensive models that encapsulate the intricate interplay of these multifaceted elements.

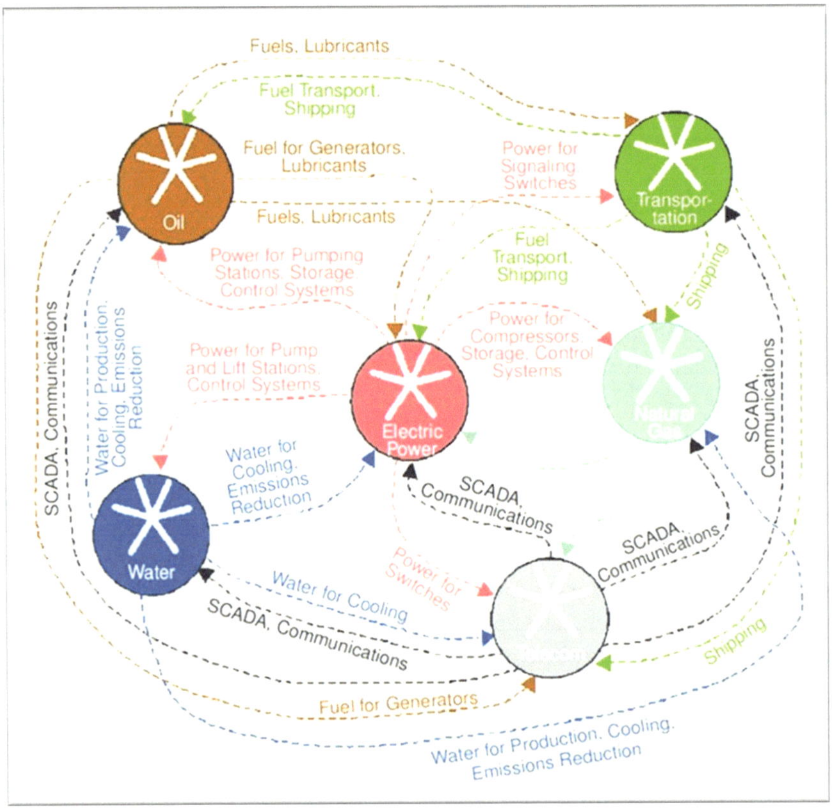

Figure 3.
Interdependence among critical infrastructures showing oil and gas sectors after Rinaldi [12].

One of the fundamental hurdles lies in the acquisition and utilization of data. The reliability, availability, and heterogeneity of data from disparate sources present a formidable challenge in constructing accurate models for interdependency assessment. Moreover, the proprietary nature of specific data within the industry often hampers the holistic analysis required for a comprehensive understanding of interconnections [2].

Recognizing the paramount importance of collaborative efforts, the chapter underscores the significance of data sharing and cooperation among industry stakeholders. Open and transparent data-sharing practices enhance the quality and comprehensiveness of available data and foster a collective understanding of interdependencies across the sector. Collaboration among industry players, researchers, and regulatory bodies catalyzes enriching methodologies, refining models, and collectively addressing the challenges associated with assessing and mitigating interdependencies (**Figure 3**) [12].

3. Case studies

3.1 Data

It is important to note that the following case studies are presented for illustrative and educational purposes. Due to the hypothetical nature of these scenarios and the unavailability of specific data, the cases are crafted to emphasize the application of interdependency analysis tools and methodologies in addressing potential disruptions. Without real-world data, these cases aim to showcase the strategic approaches organizations could adopt to enhance resilience and mitigate risks. The scenarios do not reflect actual incidents but are designed to highlight the proactive measures and decision-making processes that can be implemented in response to hypothetical challenges. The emphasis is on the conceptual application of interdependency analysis in enhancing preparedness, optimizing resource allocation, and fostering resilience within critical infrastructure operations.

3.2 Case study 1: mitigating disruptions through interdependency analysis in oil refinery operations

3.2.1 Scenario

A major oil refinery, reliant on various interconnected systems for its operations, faces a sudden disruption in its supply chain due to a pipeline rupture that hampers the transportation of crude oil to the refinery.

3.2.2 Interdependency analysis approach

Utilizing interdependency analysis tools and methodologies, the refinery management conducts a comprehensive assessment of the interconnected systems within the facility. This analysis encompasses the interdependencies between the refinery's operations, including the processing units, storage facilities, transportation networks, and supply chains. Network analysis and system modeling are employed to visualize the critical nodes, dependencies, and potential impacts of disruptions across the refinery's infrastructure.

3.2.3 Outcome and actions

Through this analysis, the management identifies critical dependencies on the affected pipeline for the crude oil supply. They uncover potential bottlenecks and vulnerabilities within the supply chain, enabling them to strategize effectively. With this insight, the refinery initiates proactive measures, including rerouting supplies through alternate pipelines, optimizing storage capacities, and implementing contingency plans to secure alternative sourcing options during emergencies. Additionally, they enhance collaboration with pipeline operators and establish emergency protocols for swift response and communication in case of future disruptions. As a result, the refinery strengthens its emergency preparedness, mitigates risks associated with supply chain disruptions, and optimizes resource allocation strategies to ensure sustained operations during unforeseen challenges.

3.3 Case study 2: enhancing resilience in offshore drilling operations

3.3.1 Scenario

An offshore drilling company operates multiple rigs in the Gulf of Mexico, prone to hurricanes and severe weather conditions during certain seasons. The company faces the constant risk of potential disruptions to its operations due to the impact of these extreme weather events.

3.3.2 Interdependency analysis approach

Recognizing the vulnerabilities inherent in offshore drilling operations, the company conducts a detailed interdependency analysis focused on its rigs, logistical support, supply vessels, and personnel transportation systems. Leveraging historical weather data, geographic information, and operational logs, the analysis employs system modeling and scenario simulations to visualize how hurricanes or extreme weather events might affect the interconnected infrastructure elements across various rigs.

3.3.3 Outcome and actions

The interdependency analysis reveals critical dependencies within the offshore drilling network. Supply vessels emerge as pivotal components for ongoing drilling operations, ensuring timely equipment, supplies, and personnel delivery to and from the rigs. Moreover, the analysis identifies potential vulnerabilities, such as disruptions to supply vessel operations due to inclement weather, which could halt ongoing drilling activities and compromise personnel safety.

To mitigate these risks, the company implements a multifaceted strategy. Firstly, they revise operational protocols, incorporating flexible scheduling and contingency plans to account for potential weather-related disruptions. Secondly, proactive measures are taken to pre-position essential supplies and equipment on rigs and supply vessels, ensuring readiness for immediate deployment in case of impending storms. Additionally, evacuation procedures are refined and personnel receive specialized training to swiftly and safely evacuate in the event of severe weather warnings.

Furthermore, the company invests in advanced monitoring systems and meteorological forecasting tools to continuously track weather patterns and anticipate potential threats well in advance. Enhanced communication systems are established to ensure seamless coordination among rigs, supply vessels, and onshore management during emergencies, allowing for swift decision-making and coordinated actions.

By understanding the complex interdependencies and their potential impact on offshore drilling operations, the company fortifies its emergency preparedness and allocates resources effectively. This proactive approach not only enhances personnel's safety but also safeguards operations' continuity, ensuring resilience and minimizing downtime during adverse weather conditions in the Gulf of Mexico.

4. Discussion

The assessment and analysis of interdependencies within oil and gas infrastructure offer critical insights that underline the imperative of proactively managing and fortifying the resilience of these essential facilities. Understanding these interdependencies not only aids in comprehending the complexity of the interconnected systems but also serves as a cornerstone for mitigating risks and ensuring uninterrupted operations, especially during unforeseen emergencies and disruptions [13].

4.1 Comprehensive understanding of interconnected systems

Interdependency analysis provides a holistic view of the intricate relationships between components such as pipelines, refineries, drilling operations, and transportation networks. This understanding allows stakeholders to grasp the dependencies and vulnerabilities inherent in the infrastructure, enabling them to identify critical nodes and pathways that are pivotal for sustained operations [14].

4.2 Risk mitigation and contingency planning

Stakeholders can proactively devise and implement robust risk mitigation strategies by recognizing the critical dependencies and potential vulnerabilities within the interconnected network. This includes developing contingency plans, establishing redundancies, and fortifying essential points to reduce the impact of disruptions. It enables companies to respond to emergencies and deploy resources efficiently and swiftly during crises.

4.3 Enhanced emergency preparedness

Interdependency analysis aids in preparing for unforeseen emergencies by enabling organizations to anticipate potential cascading effects resulting from disruptions in one infrastructure segment. This proactive approach allows for refining emergency protocols, allocating necessary resources, and developing response strategies specific to different types of disruptions [15].

4.4 Optimized resource allocation and resilience building

Insights gained from analyzing interdependencies guide efficient resource allocation, allowing companies to prioritize investments in strengthening critical infrastructure components. By investing in resilient technologies, enhancing communication systems, and implementing robust monitoring mechanisms, organizations can fortify their infrastructure against potential risks, bolstering resilience [16].

4.5 Adaptability and continuity planning

Understanding interdependencies facilitates a culture of adaptability and continuous improvement within the oil and gas sector. Companies can continuously assess and refine their strategies based on changing circumstances, technological advancements, or evolving threats, ensuring a proactive stance in adapting to new challenges and maintaining operational continuity [17].

4.6 Collaboration and knowledge sharing

Interdependency analysis encourages collaboration and knowledge sharing among industry stakeholders. Open communication and sharing of best practices, technological innovations, and risk management strategies foster a collective effort to address common challenges, strengthening the sector's overall resilience.

Assessing and analyzing oil and gas infrastructure interdependencies are pivotal in driving proactive management and enhancing resilience. This understanding empowers stakeholders to anticipate, mitigate, and respond effectively to potential disruptions, ensuring critical facilities' continued and reliable operation even in the face of unforeseen emergencies. Adopting a proactive approach based on these insights is crucial for sustaining the stability and functionality of the oil and gas industry amidst evolving challenges and uncertainties [18, 19].

5. Conclusions

The conclusions drawn in this chapter contribute to conceptual research focused on establishing the foundations of Infrastructure Interdependency within the Oil and Gas sector. This work aims to provide a theoretical framework that exemplifies the application of GIS and Remote Sensing in this critical field, investigating the essential role of interdependencies in oil and gas infrastructure.

The chapter delves into the intricacies and significance of these interdependencies, unraveling their complexity through methodologies such as network analysis, system modeling, and geospatial visualization. Despite challenges related to data availability and visualization, the collaborative efforts of stakeholders are deemed crucial for comprehensively addressing the complexities inherent in interdependencies.

Practical case studies vividly illustrate how a profound understanding of interdependencies contributes to emergency preparedness, risk mitigation, resource optimization, and the fostering of adaptability within the industry. The insights derived from these studies underscore the necessity of proactive management practices to ensure continuous operations, even when faced with unforeseen disruptions.

The foundational nature of comprehending and managing interdependencies is emphasized as crucial for fortifying the resilience of critical oil and gas facilities. The chapter advocates for a proactive approach, highlighting the significance of informed decision-making and continuous improvement. By adeptly navigating interdependencies, stakeholders can ensure the stability and reliability of the industry amidst evolving challenges and uncertainties.

Ultimately, this chapter serves as a guide toward bolstering resilience and sustaining the vitality of the oil and gas sector, offering valuable insights for industry professionals and decision-makers in navigating the dynamic landscape of interdependencies.

Author details

Rifaat Abdalla
Department of Earth Sciences, College of Science, Sultan Qaboos University, Al-Khoudh, Oman

*Address all correspondence to: rabdalla@squ.edu.om

IntechOpen

© 2024 The Author(s). Licensee IntechOpen. This chapter is distributed under the terms of the Creative Commons Attribution License (http://creativecommons.org/licenses/by/4.0), which permits unrestricted use, distribution, and reproduction in any medium, provided the original work is properly cited.

References

[1] Su H, Zio E, Zhang J, Li X. A systematic framework of vulnerability analysis of a natural gas pipeline network. Reliability Engineering & System Safety. 2018;**175**:79-91. DOI: 10.1016/j.ress.2018.03.006

[2] Abdalla RM, Niall K. Location-Based Critical Infrastructure Interdependency (LBCII). Toronto: Defence Research and Development Canada; 2010

[3] Šarūnienė I, Martišauskas L, Krikštolaitis R, Augutis J, Setola R. Risk assessment of critical infrastructures: A methodology based on criticality of infrastructure elements. Reliability Engineering & System Safety. 2024;**243**:109797. DOI: 10.1016/j.ress.2023.109797

[4] Dell'Isola M, Ficco G, Lavalle L, Moretti L, Tofani A, Zuena F. A resilience assessment simulation tool for distribution gas networks. Journal of Natural Gas Science and Engineering. 2020;**84**:103680. DOI: 10.1016/j.jngse.2020.103680

[5] Zio E. Critical infrastructures vulnerability and risk analysis. European Journal for Security Research. 2016;**1**:97-114. DOI: 10.1007/s41125-016-0004-2

[6] Arvidsson B, Johansson J, Guldåker N. Critical infrastructure, geographical information science, and risk governance: A systematic cross-field review. Reliability Engineering & System Safety. 2021;**213**:107741. DOI: 10.1016/j.ress.2021.107741

[7] Rinaldi SM, Peerenboom JP, Kelly TK. Identifying, understanding, and analyzing critical infrastructure interdependencies. IEEE Control Systems Magazine. 2001;**21**(6):11-25. DOI: 10.1109/37.969131

[8] Nieuwenhuijs A, Luiijf E, Klaver M. We are modeling dependencies in critical infrastructures. In: Papa M, Shenoi S, editors. Critical Infrastructure Protection II. ICCIP 2008. The International Federation for Information Processing. Vol. 290. Boston, MA: Springer; 2008. DOI: 10.1007/978-0-387-88523-0_15

[9] Dixon M. Information Security in the Oil and Gas Critical Infrastructure Protection (CIP) Sectors. Discovering Identify, Exploring the Science and Magic of Identity and Access Management. 2010. Available from: http://www.discoveringidentity.com/2010/07/13/information-security-in-the-oil-and-gas-critical-infrastructure-protection-cip-sectors/ [Accessed: January 6, 2024]

[10] Katopodis T, Sfetsos A. A review of climate change impacts on oil sector critical services and suggested recommendations for industry uptake. Infrastructures. 2019;**4**(4):74. DOI: 10.3390/infrastructures4040074

[11] Zlateva P, Hadjitodorov S. An approach for analysis of critical infrastructure vulnerability to climate hazards. IOP Conference Series: Earth and Environmental Science. 2022;**1094**(1):1-8. DOI: 10.1088/1755-1315/1094/1/012004

[12] Melaku HM. Investigating potential vulnerability of critical infrastructure and way forward—Recommendations to enhance security and resilience. 19 Sep 2022. DOI: 10.21203/rs.3.rs-1892615/v1

[13] Hall J, Russell T, Verschuur J, Pant R, Robertson M, Lestang T. et al. Recent

advances in global analysis of critical infrastructure networks in a changing climate. Vienna, Austria: EGU General Assembly; 23-27 May 2022. EGU22-1484. DOI: 10.5194/egusphere-egu22-1484

[14] Nirandjan S, Koks EE, Ward PJ, et al. A spatially-explicit harmonized global dataset of critical infrastructure. Scientific Data. 2022;**9**:150. DOI: 10.1038/s41597-022-01218-4

[15] Li N, Wang F, Magoua JJ, Fang D. Interdependent effects of critical infrastructure systems under different types of disruptions. International Journal of Disaster Risk Reduction. 2022;**81**. DOI: 10.1016/j.ijdrr.2022.103266

[16] Mitsova D. Integrative interdisciplinary approaches to critical infrastructure interdependency analysis. Risk Analysis. 2021;**41**:1111-1117. DOI: 10.1111/RISA.13129

[17] Sun W, Bocchini P, Davison BD. Overview of interdependency models of critical infrastructure for resilience assessment. Natural Hazards Review. 2022;**23**(1):04021058. DOI: 10.1061/(ASCE)NH.1527-6996.0000535

[18] Rachunok B, Nateghi R. Interdependent infrastructure system risk and resilience to natural hazards. In: Romeijn HE, Schaefer A, Thomas R, editors. Proceedings of the 2019 IISE Annual Conference; 2019. pp. 1-6

[19] Amelia T. The Understanding of Vulnerability of Critical Infrastructure. Critical Information Infrastructures Security Lecture Notes in Computer Science. Switzerland: Springer Nature; 2023:241-248

Chapter 9

The System's Sensitivity and Its Use in Petroleum Research

Calvin Kwesi Gafrey, Robert Wilson and Samuel Tetteh-Abaku

Abstract

Optical spectroscopic approaches, which use radiation to analyze the effects of the source radiation and its interaction with the material, are often preferred since many processes in the optical spectroscopy domain include non-ionizing radiation. They are also easier to set up, take up less space, and require less expensive apparatus. The study of reflection, absorption, transmission, and scattering qualities in optical spectroscopy reveals information on a material's molecular makeup, crystal structure, scattering coefficient, and absorption coefficient, among other characteristics. The molecules' electronic transitions, vibrational and rotational energies provide information used in optical spectroscopy investigations. Among the many other pertinent techniques already stated, petroleum exploration researchers have used X-ray fluorescence and laser-induced fluorescence techniques. These easy-to-use spectrochemical techniques can be utilized for environmental pollution evaluation and quality assessment if no other sophisticated methods are available for the analysis and research of crude oils and petroleum products.

Keywords: laser-induced fluorescence, crude oil, refined fuels, principal component analysis, fluorophore and physicochemical parameters

1. Introduction

Depending on how much hydrogen is present in it, petroleum can be classified as a paraffin, intermediate, or naphthenic base. The exact makeup of petroleum's hydrocarbon and non-hydrocarbon components is presented in **Tables 1** and **2**, respectively. It is the main energy source for modern civilization, which currently supplies 85% of the world's energy needs, and this trend is expected to continue [1]. It takes millions of years for crude oil to develop and then be extracted. You may find these natural oils all around the world. They are obtainable from a wide variety of other geological sources and include intricate trace elements that define the geological makeup of the area and, consequently, the oil field [2].

Crude oil composites consist of non-hydrocarbon components such as sulfur (S), nitrogen (N), oxygen (O), trace elements, and salts as well as aliphatic, aromatic, high-molecular-weight organic molecules [3]. Trace elements, including iron, zinc, cobalt, nickel, sodium, and magnesium, characterize the characteristics of crude oil and the host sediment [4, 5]. The kind of technology to be utilized for

Hydrogen family	Characteristics	Major hydrocarbons	Observations
Paraffins	Straight carbon chain	Methane, Ethane, Propane, Butane, Pentane, Hexane	General formula C_nH_{2n+2} The boiling point increases as the number of carbon atoms increases. With the number of carbon atoms 25–40, paraffin becomes waxy.
Isoparaffins	Branched carbon chain	Isobutane, Isopentane, Neopentane, Isooctane	N/A
Olefins	One pair of atoms of carbon	Ethylene, Propylene	General formula C_nH_{2n} In crude oil, olefins are not present but are produced during the process. Olefins with low molecular weight have excellent anti-knock properties.
Naphthenic compounds	Five or six atoms of carbon in the ring	Cyclopentane, Methyl cyclopentane, Dimethylcyclopentane, Cyclohexane, 1,2-Dimethylcyclohexane	General formula $C_nH_{2n+2} - 2R_n$ R_N is the number of naphthenic rings. By weight, the average crude oil contains around 50% naphthene. Naphthenes are modestly good.
Aromatic compounds	Six carbon atoms in the ring with three around the linkage	Benzene, Toluene, Xylene, Ethyl Benzene, Cumene, Naphthalene	In kerosene and lubricating oil, aromatic compounds are not desirable.

Table 1.
Petroleum hydrocarbon composition [3].

Non-hydrocarbons	Products	Observations
Compounds of sulfur	Hydrogen sulfide	Unwanted due to foul odor (0.5 to 7%)
Compounds of nitrogen	Quinoline Pyridine Pyrrole Indole	After exposure to sunlight, the presence of nitrogen compounds in gasoline and kerosene degrades the color of the fuel. They can cause gum formation. Less than 0.2%, usually.
Compounds of oxygen	Naphthenic Phenols and acids	These acids lead to corrosion and contamination problems at various stages of the manufacturing process. Materials trace amount to no more than 2%.

Table 2.
Non-hydrocarbon composition of petroleum [3].

drilling and refining depends on their qualities. Oil wells and different oil fields may have varied compositions of crude oils because of their diverse geological origins [6].

Understanding the properties of crude oil from a specific geological source is one of the most important factors in drilling optimization. Furthermore, the features offer accurate details regarding the type of refinery technology required for processing and environmental safety requirements [5]. Without adequate data analysis to identify the right technology to target the ideal drill location, a great deal of time, money, and environmental damage would be incurred [4, 5]. Therefore, the petroleum industry places great importance on the development of scientific processes and procedures for determining the characteristics of distinct crude petroleum oils from different geological sources.

Studies have demonstrated that the quality of crude oil is ascertained chemically using techniques like the Grote-Krekeler, Herman, and Mortiz procedures, among other non-spectroscopic techniques. To get the desired findings, they must heat the samples for several hours at temperatures as high as 750°C [7]. These procedures are not only time-consuming and somewhat costly, but they also damage the samples by requiring many steps.

The ease of use and sample management in spectroscopy have facilitated the creation of novel spectroscopic methods for figuring out the physicochemical characteristics of petroleum products. These methods have become more relevant in the examination of crude oil because they have a distinct diagnostic process [8]. In petroleum technology, non-destructive spectroscopic methods like Gamma Spectroscopy (GS), X-ray Fluorescence (XRF), and Laser-Induced Fluorescence (LIF) are most frequently employed for the examination of crude oil [6]. The rationale for using these methods is that they provide quick responses, display excellent sensitivity and selectivity, and include useful details on the inherent chemical properties of any sample they analyze [6].

Scientists now prefer to employ laser-induced fluorescence, X-ray fluorescence, and gamma spectroscopy techniques because they offer more precision and accuracy when identifying main and minor elements and minerals. Spectral line interference is not very prevalent, and sample preparation is non-destructive. The specimen form can be paste, liquid, gas, solid, or powder. A standard wavelength fingerprint for crude oils can be created using the fluorescence spectrum profile of the light and heavy oils obtained by laser-induced fluorescence [8].

It is possible to determine a substance's elemental makeup using X-ray fluorescence. Compared to techniques like Instrumental Neutron Activation Analysis (INAA), which was employed to characterize the Jubilee and Saltpond crude oils, it is preferable due to its ease of sample handling [6]. Despite INAA's ability to assess sulfur contents, sample preparation using standards presented a challenge. Compared to X-ray fluorescence spectroscopy, which employs computational techniques to precisely identify the sulfur components in crude oil, this method is comparatively costly and laborious. Natural oils with high sulfur content release dangerously high amounts of sulfur dioxide (SO_2), which are harmful to both human health and the environment. Suffocation, increased breathing rates, and lung inflammation are among the respiratory effects of sulfur dioxide [6, 9, 10]. The worsening of chronic bronchitis, asthma, and eye and throat irritation is the most detrimental health impact. It plays a part in the creation of acid rain, which seriously damages materials, aquatic and terrestrial ecosystems, as well as the human population [11].

2. Non-hydrocarbons

The amount of knowledge on the composition of petroleum has expanded because of newly invented tools and methods. Researchers can identify the class of

Element	Composition (%)
Carbon	38–87
Hydrogen	10–14
Nitrogen	0.1–2
Oxygen	0.05–1.5
Sulfur	0.05–6
Metals	< 0.1

Table 3.
Elemental percentage range [13].

hydrocarbon structures as well as the composition of all hydrocarbons and their structures [3, 12]. Examples of these new techniques include nuclear magnetic and paramagnetic resonances. These techniques offer information about individual hydrocarbon composition. The discovery and recovery of trace elements from crude oil are technically complex and have yet to be regularly used, although scientific experimentation is in progress [11]. **Table 3** displays the elemental percentage range showing element versus composition by percentage.

Any specific hydrocarbon species' physical and chemical characteristics depend on the kind and quantity of chemistry between its carbon atoms and the amount of carbon atoms in the molecule [14]. The yield and quality of refined petroleum products are determined by the various proportions of the enormous hydrocarbon class and distribution, not to mention the percentage composition of single and other multi-elements and their impact on crude oil. The refined fuels' market price assessment fluctuates based on the current average price of all crude oils [6]. The yield quality of refined petroleum is determined by classifying it into lighter and heavier grades. Lighter grades offer quality yields, while heavier grades require a carbonization procedure.

Petroleum quality and sulfur concentration are related. Crude oil's value on the global market is primarily determined by its American Petroleum Institute (API) gravity, with higher gravity oils fetching higher prices. Crude oils with low sulfur concentrations are considered sweet on a global scale. The range of sulfur contents for sweet and sour crude is around 0.1–0.5% and 1–3.3%, respectively [11, 13]. Heavy-grade and sour crude oils are less popular on the market than light and sweet crude oils because they need more work to process. Because of their low API gravity, heavy oils require more time, money, and energy to refine, making light and sweet crude oil the most profitable [3].

3. Measurement techniques used in petroleum research

Optical spectroscopic approaches, which use radiation to analyze the effects of the source radiation and its interaction with the material, are often preferred since many processes in the optical spectroscopy domain include non-ionizing radiation. Additionally, they are easier to set up, consume less space, and have inexpensive instrumentation [15–17]. The study of reflection, absorption, transmission, and scattering qualities in optical spectroscopy reveals information on a material's molecular makeup, crystal structure, scattering coefficient, and absorption coefficient, among other characteristics. The electronic transitions, vibrational and rotational energies of the molecules provide information that is used in optical spectroscopy investigations [13, 17, 18].

The aforementioned parameters are ascertained using two analytical methods. These are molecular spectroscopy and elemental or atomic spectroscopy. Among the frequently used measurement methods in molecular spectroscopy are fluorescence spectroscopy (FS), absorption spectroscopy, Fourier transform infrared spectroscopy (FTIR), and Raman spectroscopy (RS). An optical method for analyzing emissions brought on by photons being absorbed by an atom is called fluorescence spectroscopy [10, 13, 19]. Atomic electronic transitions are the source of fluorescence emission, which happens in the electromagnetic spectrum's optical range between 300 and 700 nm. This method's sensitivity and specificity make it appropriate for molecular analysis [17, 20]. Both liquids and solids have wide emission bands of fluorescence. Excited species typically de-excite in a few nanoseconds to microseconds after absorbing energy and releasing light at a wavelength that is longer than that of the excitation. A photomultiplier tube (PMT) or filtered photodiodes are used to record this fluorescent light [21].

4. The process of fluorescence

Luminescence, the term for the emission of light from any substance, is the result of electrically excited states. Depending on the type of excited state, luminescence is technically divided into two groups: fluorescence and phosphorescence [20]. When stimulating radiation persists, a substance that has absorbed energy releases it as light, a process known as fluorescence [22]. When photons are emitted, the second electron in the ground-state orbital is paired with the excited singlet state electron in the excited orbital, and the electrons quickly reach the spin-allowed ground state [20]. Fluorescence emission rates are approximately $10^8\ s^{-1}$ on average, meaning that a typical fluorescence lifetime is close to 10 ns (nanoseconds) $(10 \times 10^{-9}\ s)$ [17, 20].

5. The fluorescence emission theory

Lakowicz [20] asserts that the fluorescence lifespan and quantum yield of a fluorophore, a fluorescent substance, are its most crucial properties. The lifetime establishes how long the fluorophore has to interact or diffuse in its surroundings, whereas the quantum yield is the ratio of photons released to photons absorbed. The quantum yield is determined by the fraction of fluorophores that decay by emission, which is expressed as

$$Q = \frac{\Gamma}{\Gamma + k_{nr}} \tag{1}$$

where (Γ) denotes the fluorophore's emission rate and non-radiative decay rate $S_o(k_{nr})$ [13, 20].

The lifespan of fluorescence is close to 10 ns. The lifetime expression is provided by the equation;

$$\tau = \frac{1}{\Gamma + k_{nr}} \quad (2)$$

The fluorescence process is depicted in **Figure 1** through a condensed Jablonski diagram. Since not all molecules release photons at exactly t = τ, the lifetime is an average of the amount of time spent in the excited state. Known as the intrinsic or natural lifetime, it is the fluorophore's lifespan in the absence of a non-radiative process and is determined by the equation;

$$\tau_n = \frac{1}{\Gamma} \quad (3)$$

where τ_n is the natural lifetime.

The radiative decay Γ can be calculated using

$$\begin{aligned}\Gamma &= 2.88 \times 10^{-9}\, n^2 \frac{\int F(\overline{v})d\overline{v}}{\int F(\overline{v})d\overline{v}/\overline{v}^3} \int \frac{\varepsilon(\overline{v})}{\overline{v}}\, d\overline{v} \\ &= 2.88 \times 10^{-9}\, n^2 < \overline{v}^{-3} >^{-1} \int \frac{\varepsilon(\overline{v})d\overline{v}}{\overline{v}}\end{aligned} \quad (4)$$

where ε ($\overline{v}$) is the absorption spectrum, n is the medium's refractive index, and F ($\overline{v}$) is the emission spectrum displayed on the wavenumber (cm^{-1}) scale [13, 20]. It works rather well, especially for polynuclear aromatic hydrocarbons, and the integrals are calculated over the $S_o \leftrightarrow S_1$ absorption and emission spectra. For a variety of uses, fluorescence spectroscopy has developed into multiple formats. These include, among other methods, Fourier Transform Infrared Spectrometry (FTIR), Time-Resolved Laser-Induced Fluorescence (TRLIF), Rapid Optical Screening Tool (ROST), Temporal Fluorescence Alteration (TFA) analysis, and Laser-Induced Fluorescence (LIF) [13].

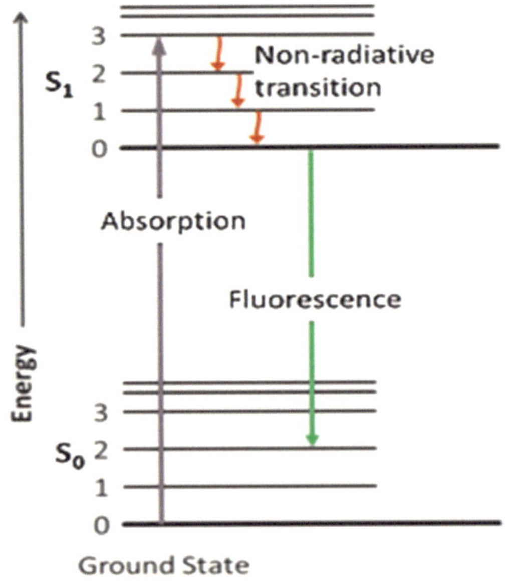

Figure 1.
A condensed Jablonski diagram that illustrates the fluorescence procedure.

Beyond its high sensitivity and excellent selectivity, LIF is unquestionably a vital instrument for environmental monitoring [13, 23, 24]. Since soil and water contamination poses a serious threat to the ecosystem, it is necessary to conduct an in-depth study using low-cost methods [25]. Compared to non-spectroscopic approaches, using LIF for the study and analysis of refined fuels offers significant advantages in terms of sample size, analytical precision, ease of sample handling, and potential *in situ* measurements [8, 13]. The reaction of aromatic molecules with electromagnetic radiation produces fluorescence emission from refined petroleum.

The physicochemical characteristics of the studied samples impact the fluorescence spectra and their intensities [8, 26]. Therefore, evaluating the quality of petroleum-related goods using LIF as a quick and straightforward spectrochemical instrument is conceivable. The fluorescence spectra and their intensities are affected by the physicochemical properties of the samples under investigation [8, 26]. Therefore, it is plausible to use LIF as a fast and straightforward spectrochemical tool in assessing the quality of petroleum-related products.

The most popular measurement method that organic petrographers use to characterize sedimentary organic materials is called temporal fluorescence alteration (TFA) analysis [27]. This phenomenon mainly involves irreversible and frequently stunning changes in color and intensity, which happen when fluorescent macerals are continuously exposed to ultraviolet (UV) radiation. The petrochemical base of petroleum samples has been examined and understood by applying the TFA analysis. Organic petrographers now understand that the maturity and type of irradiation maceral are the primary determinants of the kinetics and amplitude of this occurrence, thanks to these observations examined in the TFA analysis [27].

For hazardous waste sites, the Rapid Optical Screening Tool (ROST) offers real-time field screening of the chemical and physical properties of soil contamination caused by petroleum [28]. A cone penetrometer probe (cone penetrometer tube (CPT)) coupled to a laser-induced fluorescence sensor makes up the ROST. Any traditional CPT system with a sapphire window installed outside the stainless-steel probe and above the cone penetrometer tip can be used with the ROST system. The ROST sensor is a trustworthy field screening tool because light from an excited laser travels through the sapphire tip and fluoresces, providing information on the aromatic pollutants in the soil [29].

Petroleum research and exploration employ several potent optical techniques, one of which is Time-Resolved Laser-Induced Fluorescence (TRLIF). TRLIF is a unique method for detecting fluorescent metal ions in soil, petroleum products, and tainted marine samples [30], in addition to being a practical method to detect oil contamination in water up to a concentration of 0.5 ppm (parts per million). In the presence of silicic acid (Si(OH)-4), TRLIF in conjunction with parallel factor analysis (PARAFAC) showed its special usefulness in the identification of uranyl ($UO2_{2+}$) [13, 31, 32]. Determining the elemental concentrations in crude oil samples, refined fuels and petroleum gases may not be possible, particularly if the method does not reveal the levels of concentration in the oils [13, 30].

6. Inner electron spectroscopy

High-energy electromagnetic waves known as X-rays are created when quickly moving electrons come to a stop. The characteristic component and the Bremsstrahlung make up the radiation that is produced [13, 17]. Two types of X-rays are produced during the process of X-ray emission: continuous X-ray radiation and distinctive X-ray radiation. When high-energy electrons in an X-ray tube are decelerated by the electric field of a metal target, continuous X-ray radiation is created [13, 17]. This procedure produces an X-ray energy spectrum that is continuous, without any noticeable peaks, and ranges from low to high. On the other hand, when outer-shell electrons transition to replace the vacancies left by inner-shell electrons being expelled from atoms in the target material, characteristic X-ray radiation is created. As a result, X-rays with certain energies that correlate to the energy differential between the implicated electron energy levels are released. These energies show up in the X-ray spectrum as discrete peaks that are indicative of particular elements [13, 17].

When charged electrons are attacked, they decelerate and alter their direction of motion, which excites the atom to a higher energy level and produces the Bremsstrahlung (continuous part). As seen in **Figure 2**, an electron hole is seen to travel out of the other shells to cause the transition to a lower energy state. The consecutive atomic energy losses are obtained as a sequence of emission lines. The migration of the electron hole from the L-shell to the M-shell is represented by the transition, which is shown by $K\text{-}\alpha$, $K\text{-}\beta$, $K\text{-}\gamma$ [13, 17].

The discovery of X-rays and electromagnetic radiation with a wavelength similar to the size of separation between crystal planes (0.1 nm) was made by Wilhelm C. Röntgen. This was looked at by Max von Laue, William Henry Bragg, and William Lawrence Bragg in their X-ray-based analysis of crystal structures. This was a significant turning point in the history of X-ray crystallography. Atomic X-ray studies have produced an extremely accurate mapping of inner electrons' energy levels, and many emission line wavelengths have now been determined [17]. The energy states of the inner electron shells are relatively less affected by the atoms' chemical surroundings. Therefore, a spectral analysis of the distinctive X-ray emission is a good choice for elemental analysis. Moseley's law provides the relationship between an X-ray line's wavelength and the nuclear charge Z of the associated atom;

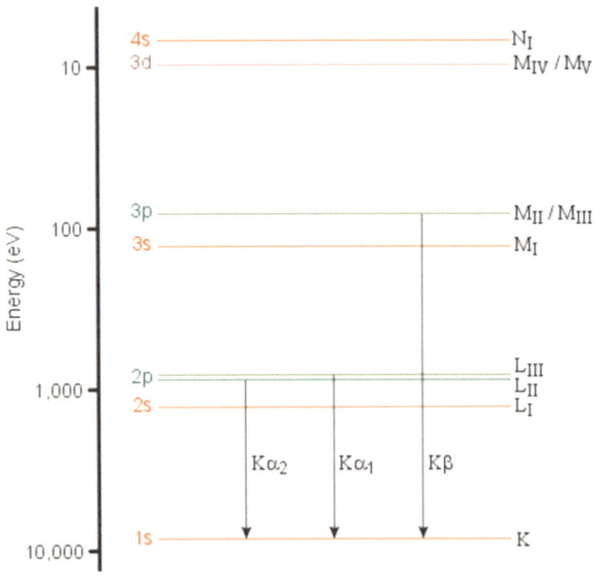

Figure 2.
X-ray transition of inner electron energy levels.

$$1/\sqrt{\lambda} = C(Z - \sigma) \tag{5}$$

According to the law, the square root of the X-ray emission frequency is related to the atomic number. A particular spectral species is characterized by constants C and σ [17]. Bohr's fundamental atomic model can be used to derive the law of Moseley. **Figure 3** displays a Moseley diagram for K_α and L_α emission. It is possible to identify components in a sample by using an illustration like this one.

7. X-ray emission spectroscopy

One of the traditional inner-shell photoelectric effect methods is X-ray emission, which can be created by employing energetic electrons or heavier charged particles. It is possible to observe fluorescence by creating an inner-shell vacancy in the material by X-ray radiation. As long as the stimulating radiation is present, a portion of the energy a substance absorbs is released as light, by a process known as fluorescence [13, 33].

8. Energy-dispersive spectroscopy

It is efficient to process X-rays with an energy-dispersive device. This method does not require any crystal examination because the radiation hits an energy-dispersive detector straightaway. The energy of the X-ray quanta is employed to generate voltage pulses via a silicon lithium-doped detector (Si (Li)). An Analog-to-Digital Converter (ADC) in a multi-channel analyzer (MCA) is used to separate the pulses based on their energy (amplitude) [17]. On a system display, the entire spectrum gradually becomes visible. The linewidth is derived from a 150 eV silicon lithium-doped

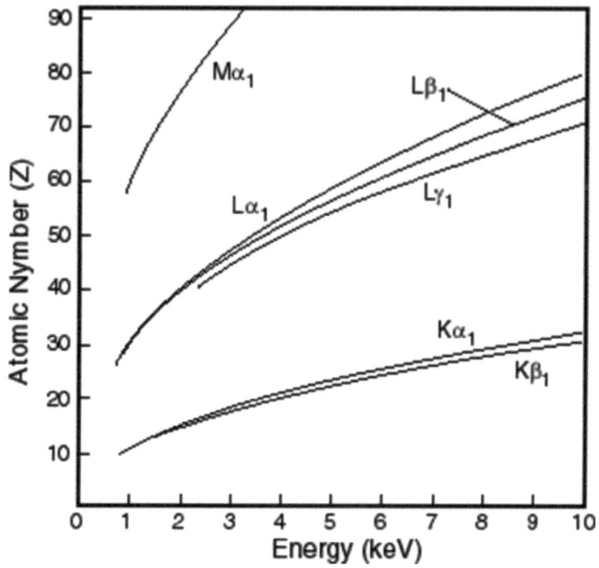

Figure 3.
Diagram that demonstrates Moseley's Law.

detector (Si (Li)) detector. It is possible to perform multi-element analysis using the energy-dispersive system [13, 17].

9. The sensitivity of the system and its application in petroleum research

The signal of a standard sample divided by the root mean square noise level is a common way to express a system's sensitivity. This makes it possible to compare the various methods and approaches employed in experimental research across laboratories worldwide. Researchers have increasingly turned to light and light-based spectroscopic techniques to examine sample characteristics and explore material qualities [10, 34, 35]. Because they are non-destructive, spectroscopic techniques frequently involve little to no sample preparation and offer quick analysis [10, 34, 35].

The fuel production process has adjusted to a global trend of reduced toxic chemicals in combustion products [34]. Numerous standardized processes can be used for different fuel classes when determining the elemental composition of crude oils and petroleum fuels. For this reason, an analysis method like X-ray fluorescence is necessary. Two recommended methods are used to assess the sulfur concentration in diesel oils, aviation fuels, and crude oils. The Slovak Technical Standard, like the European Standard, uses Wickbold's method to determine elemental composition and Grote Krekeler's method (ASTM D 1551), which proved uncomfortable.

Since practically every element in the periodic table can be found in crude petroleum oil, determining and measuring the concentration of element composition in crude is essential and of interest [9, 10, 13]. Information on these components is important and vital for drilling, environmental contamination assessment, and oil field development [6, 9, 10, 13].

10. Experimental setup of laser-induced fluorescence (LIF) measurement

The components of this spectral research or analysis setup are an analyte, a detecting equipment, and a radiation source. The radiation source is a 405-nm wavelength continuous-wave (CW) diode laser with a maximum output power of 100 mW (milliwatt). It has a laser diode driver with transistor-transistor logic (TTL) modulation of 0–20 kHz that requires an output voltage and current of 12 V and 2 A, respectively. Optical elements were employed, including a fiber port micro-positioner (PAF-SMA-5-B, Thorlabs, USA) and a bifurcated optical fiber probe (Ocean Optics, USA). When the optical fiber cable is in use, it functions as an entrance slit through which light enters. The bifurcated optical fiber, which has a core diameter of 600 μm, was linked to the fiber port positioner, which directed the beam out of the microscope objective lens (CP09/M, Thorlabs, USA).

The fluorescence of the crude oil and refined petroleum samples was detected using a detection system consisting of a 450-nm cut-off long-pass filter and a CCD-based USB2000 spectrometer (USB2000, Ocean Optics). This spectrometer responds to photon activity between 200 and 1100 nm and runs on low power characteristics of 100 mA at 5 V. An asymmetric crossed Czerny-Turner optical bench disperses the light onto a 2048-element linear silicon CCD array detector. The CCD detector's data are obtained by selecting an integration time between 1 and 60 ms. The gathered data are converted into digital form and sent to the Ocean Optics USB2000 spectrometer's 001Base32 operating software on the personal computer (PC).

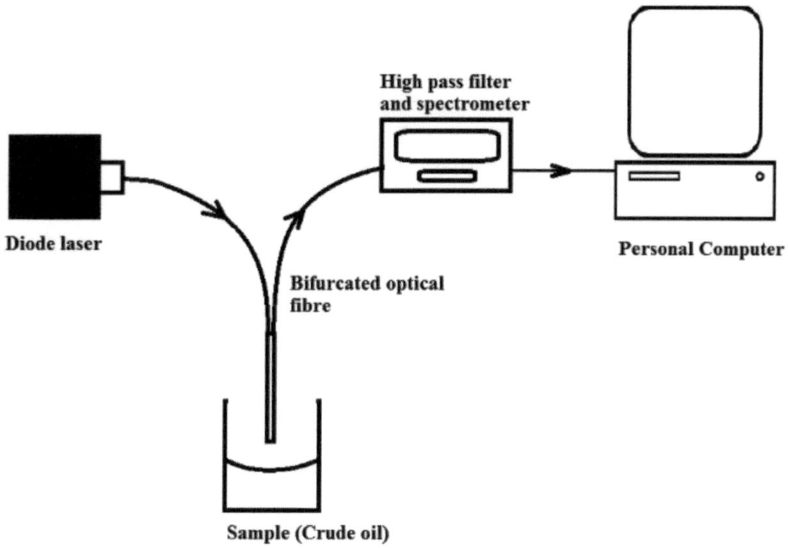

Figure 4.
Diagrammatic schematic of the LIF measurement apparatus displaying the analyte (crude oil), the radiation source (diode laser), and optical parts (Fiber Probe) and the detection method (Spectrometer) using a filter.

The operating software for the 001Base32 spectrometer is a 32-bit program that can be customized by the user for advanced retrieval and display applications. The real-time measuring display enables the user to make choices and modifications. A significant amount of data can be automatically saved and transferred to other software platforms for additional analysis. The program is computer-controlled for data collection (**Figure 4**).

11. Laser-induced fluorescence (LIF) measurement

At the Laser and Fiber Optics Centre (LAFOC), Department of Physics, University of Cape Coast (UCC), the fluorescence intensities of four crude oil samples were measured using a laser source. Each of the four samples had five replicates, labeled JF, TF, SF 1, and SF 2. Every sample was exposed to a 405-nm CW diode laser. The fluorescence that was released was collected and passed through an optical fiber that was placed behind a converging lens 90 degrees away from the sample. The fiber was then sent through a long-pass filter and onto the USB2000 spectrometer (Ocean Optics, USA), where the spectrum was shown on a computer screen. After the spectrum was obtained, the data from every sample were preserved. The 001Base32 program exported the spectrum's intensity and wavelength data into MATLAB (MATLAB R2019a) for additional processing and analysis.

12. Data interpretation for LIF measurement

Each sample's fluorescence spectra were deconvolved to identify the hidden peaks that contributed to the overall spectra. To find the similarities and differences in the data collected, an unsupervised pattern recognition approach was used: Principal

Component Analysis (PCA). A classification model was created using the input variables from the PCA's Score plot.

From the four crude oil samples that were sampled from the four distinct oil wells, 600 LIF spectra were acquired. The training set consisted of 400 LIF spectra or 66.67% of the total LIF spectra that were acquired. For the verification set, 200 LIF spectra, or 33.33% of the total LIF spectra, were used.

13. Experimental setup of X-ray fluorescence measurement (XRF)

The AMP TEK X-123 Experimenter Kit (AMPTEK INC., USA) with a rapid silicon drift detector (SDD) was used in this investigation. Stainless steel 316 (SS316) test sample, collimator, and filter holder-designed Mini-X USB (universal serial bus)-controlled X-ray tube (radiation source), and XRS-FP2 quantitative analysis software package interfaced with a computer for material analysis are all part of this system. Three distinct application modes are available for it:

1. Single-layer samples (XRS-FPS BULK) for single-layer thickness and compositional analyses.

2. Samples of multiple layers of thin films for the study of composition and multiple layers' thickness (XRS-FP2 MTF).

3. Single-layer samples, utilizing an incoming electron beam rather than an X-ray beam to determine composition and single-layer thickness (XPS-FP2 EPXA). The radiation source and test stand with shielding and sample enclosure eliminate the risk of radiation leakage, as shown in **Figures 5** and **6**, respectively.

14. Scale calibration in the Amptek DPP MCA

The Amptek DPP MCA program elaborates on the scale calibration process. For the calibration, two peaks were selected, and their energies were recorded.

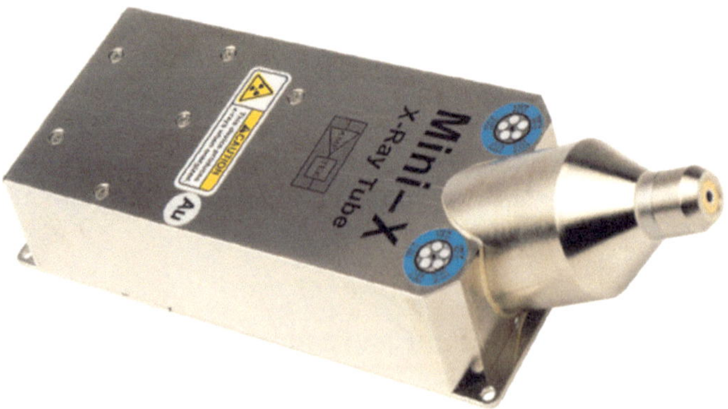

Figure 5.
In-built Mini-X-ray tube radiation source.

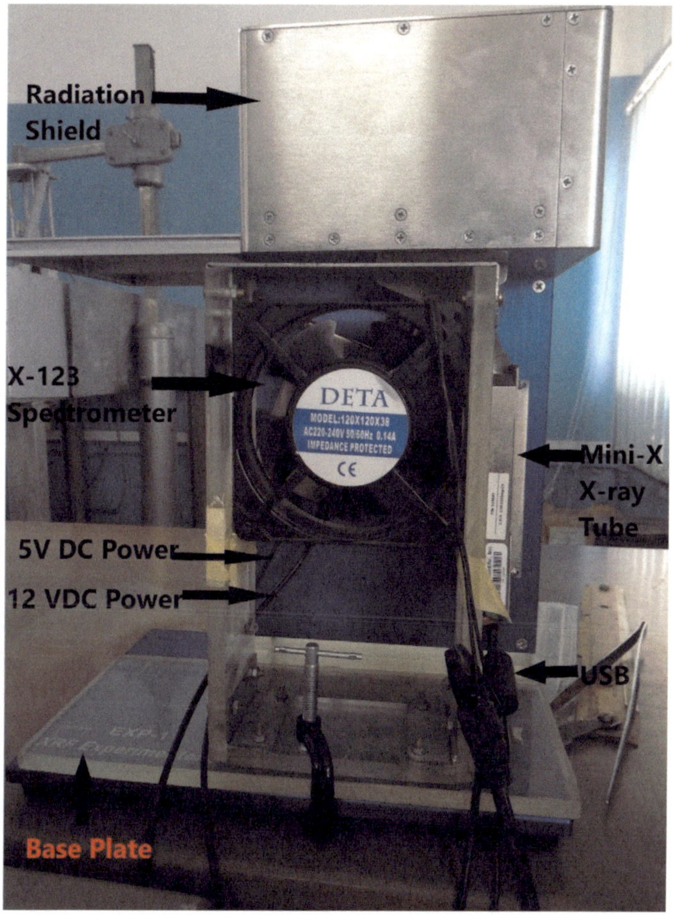

Figure 6.
Kit for AMPTEK EXP-1 X-123 XRF experimentation displaying the Mini-X-ray tube radiation source and spectrometer (X-123) both located below the container for radiation protection.

Every peak's region of interest (ROI) was indicated. Clicking the left base of the first picked peak highlighted the entire peak with the pointer while holding down the "U" key.

The calibration dialog box was accessed and moved to display the peaks using the "Calibrate" toolbar key. Information was added, and the apex was emphasized.

After adjusting the regions of interest's beginning and endpoints, the centroid button on the conversation box was pressed to enter the peak's center position in the Channel box. After taking the same actions as before for the first peak, the cursor was placed once more into the second peak. To finally calibrate the scale in kiloelectron-volt (keV), the energy units (keV) were entered into the Unit box.

15. XRF measurement

Using a calibrated Pasteur tip ejector pipette, every sample was transferred into two-milliliter plastic cups secured by clear plastic sheaths known as Mylars. The Mylars were soaked in strong acetone and then cleaned with filtered water to avoid

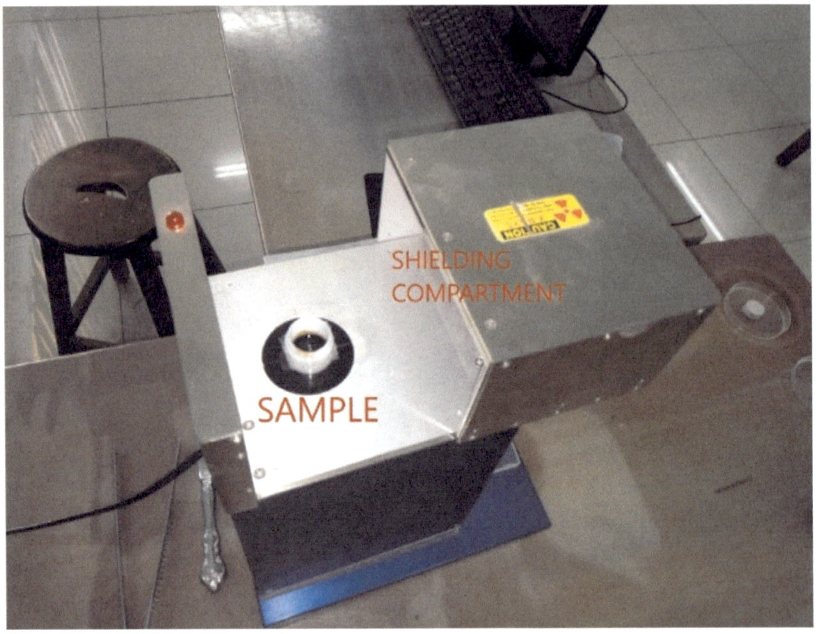

Figure 7.
Sample holder inside the AMPTEK shielding compartment EXP 1 X-123 XRF Experimenter Kit displaying the detector for the sample arrangement.

any surface contamination. To confirm the accuracy and consistency of the measurements, four replicates of each sample were examined [10, 13].

At the Ghana Atomic Energy Commission (GAEC), Kwabenya, the National Nuclear Research Institute (NNRI) facility is the location where the X-ray fluorescence measurements were conducted. The EXP-1 XRF Experimenter Kit was used to conduct radiation exposure on four sets of samples that were delivered to the Spectrometry laboratory. The system was run at 45 kV of voltage and 5 μA (microamperes) of current. The system consists of a DP5 Digital Signal Processor and an X-123 Complete X-Ray Spectrometer with an SDD detector. The Mini-X X-ray tube has a sample chamber, a radiation-shielding container with safety interlocks, a collimator, and a location to mount filters. **Figure** 7 depicts the sample holder within the shielding compartment, whereas **Figure 8** displays the shielding chamber along with its safety plunger.

16. Data analysis of XRF measurement

Amptek DPP MCA software package for spectrum analysis was used for qualitative analysis to determine constituents in the sample. Trace element identification is made simple by features like background subtraction, centroids of known regions of interest, spectra scaling, etc. The acquired data were stored in widely supported software formats, such as ASCII (American Standard Code for Information Interchange) or Amptek MCA. The bAxil software version 1.6 was used for quantitative analysis to determine the sulfur contents of the sampled crude oil samples. Without the need for

Figure 8.
The safety plunger-equipped shielding section displaying the AMPTEK EXP 1 X-123 XRF experimenter's safety features set.

standards, the software was able to calculate the elemental concentrations from the examined spectra by calculations based on the Fundamental Parameter approach [13].

17. Results and discussion

17.1 Deconvolution of fluorescence spectra in samples of crude oil

The Gaussian Function Fitting approach was used to deconvolve each sample's spectra and retrieve information about the concealed fluorescence peaks. JF, TF, SF 1, and SF 2 oil wells' deconvoluted fluorescence spectra are displayed in a composite graph (**Figure 9**). These fits had correlation coefficients (r^2 values) greater than 0.99.

The deconvoluted fluorescence spectra were made up of Gaussian components that were measured at various peak intensities and wavelengths. It is evident from **Figure 9** that the important fluorescence peaks fall between 510 and 750 nm in wavelength. These fluorescence peaks match other researchers' observations of

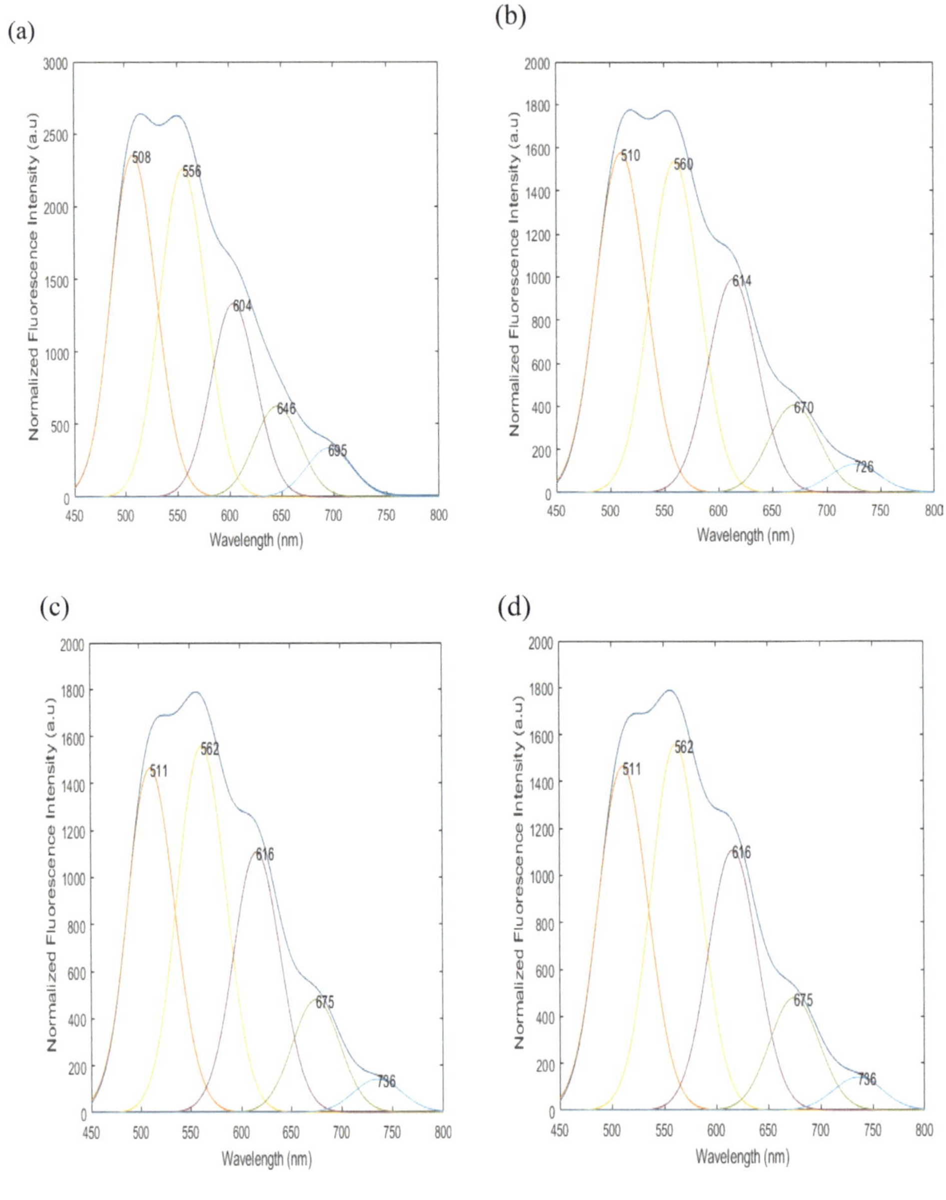

Figure 9.
Deconvoluted fluorescence spectra of (a) JF, (b) TF, (c) SF 1, and (d) SF 2 showing the hidden peaks.

fluorescence emission results for crude oils [9, 13]. Given that they are both drilled in the same field, the peak wavelength similarities between SF 1 and SF 2 indicate that they may share chemical features and have similar fluorophores. **Table 4** displays the deconvoluted fluorescence spectra's peak wavelengths and intensities.

The differences in the peak wavelengths suggest that the crude oil samples from oil wells JF, TF, and SF have different fluorophore compositions and do not have the same molecular characteristics [8, 9, 13]. **Table 5** displays the mean maximum wavelength values of the deconvoluted fluorescence spectra for the oil samples. For the first time, the conventional deconvoluted peak wavelength range of 510–750 nm is

	JF		*TF*		*SF 1*		*SF 2*	
Peaks	*λ (nm)*	*I (a.u.)*	*λ (nm)*	*I (a.u.)*	*λ (nm)*	*I (a.u.)*	*λ (nm)*	*I (a.u.)*
1	508	2350	510	1576	511	1465	511	1438
2	556	2257	560	1535	562	1557	562	1495
3	604	1331	614	990	616	1107	616	1075
4	645	623	670	404	675	479	675	468
5	695	335	726	130	736	139	736	138

Table 4.
Deconvoluted fluorescence spectral results for crude oil specimens from each of the four oil wells.

Peaks	Wavelength (nm)
1	510.00 ± 1.41
2	560.00 ± 2.83
3	612.50 ± 5.74
4	666.50 ± 13.87
5	723.25 ± 19.41

Data behind ± for this work are standard deviations.

Table 5.
Average peak fluorescence wavelength values of deconvoluted spectra of crude oil samples from the oil fields JF, TF, SF 1, and SF 2.

provided by these average peak wavelengths, identifying the precise peak wavelength range at which fluorescence is taking place. The peak wavelengths also imply that the fluorescence emission in the crude oil samples JF, TF, and SF is caused by five distinct fluorophores.

18. Crude oil sample fluorescence spectroscopy data: principal component analysis (PCA)

Investigating Principal Component Analysis (PCA) to identify the fluctuations in the fluorescence spectra was worthwhile to gain a deeper understanding of the distinctions between JF, TF, SF 1, and SF 2. By lowering the dimensionality and collecting the principal component coefficients and eigenvalues of the data's covariance matrix, the PCA was able to simplify the sizable data set that was acquired from the fluorescence measurements. After that, each spectrum was plotted as a single point in PC space using the variance for the complete set of data.

The scree plot displaying the eigenvalues of the covariance matrix of the crude oil sample fluorescence spectra is displayed in **Figure 10**. It chooses which principal components to preserve in a principal component analysis (PCA) or how many variables to keep in an exploratory analysis (EA).

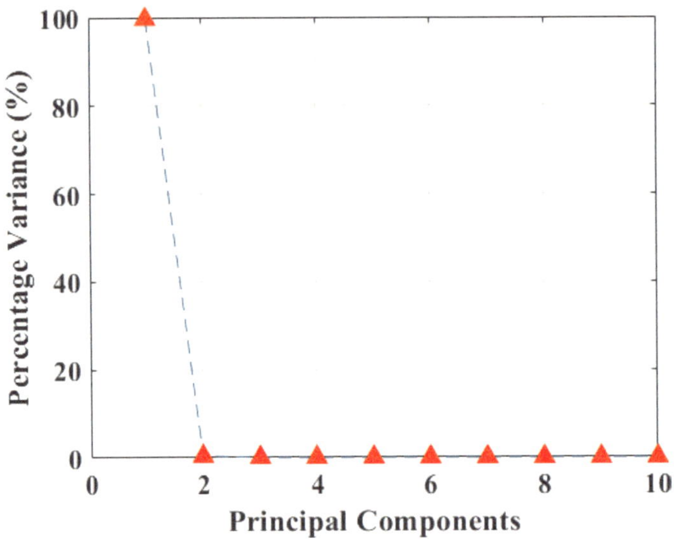

Figure 10.
The number of major components and how much each one contributes to the variation in the data set described as a percentage.

According to the amount of explained variation of the eigenvalue as well as previous research [9, 13, 36], the best potential principal components may finish at the point where the scree plot flattens and levels out. To represent all of the data, two principal components (PCs) were chosen. The maximal variance in the fluorescence spectra was maintained by the two PCs (first principal component (PC1) and second

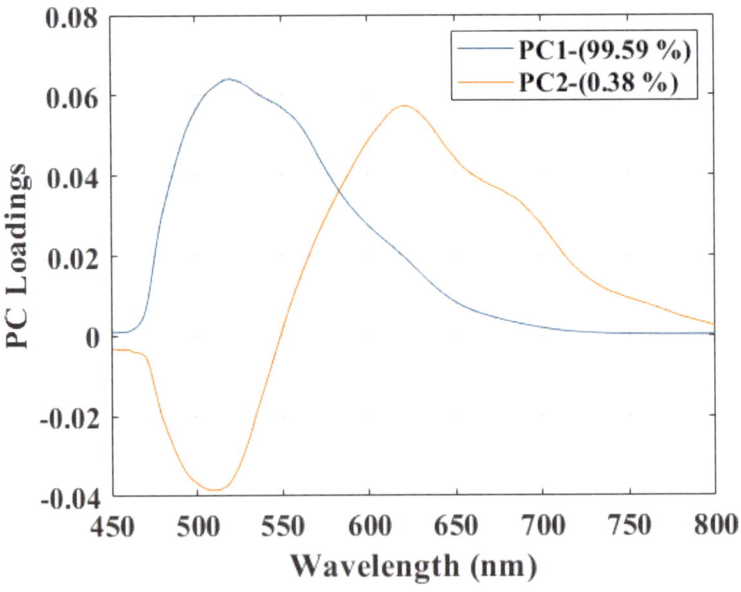

Figure 11.
Information-giving loadings plot shown from the PCA of the LIF spectrum regarding the noteworthy causes of variance found in the data.

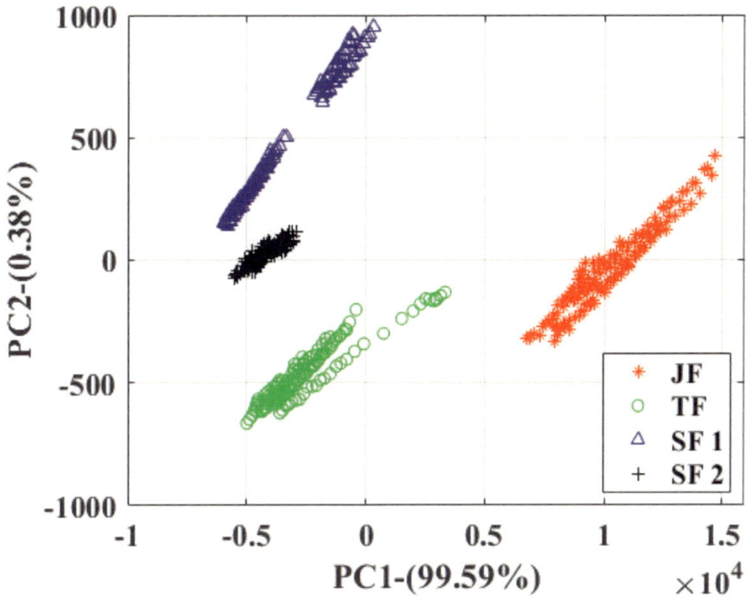

Figure 12.
Score plot from Principal Component Analysis of PCs 1 and 2; the LIF spectra of the four samples of crude oil—JF, TF, SF 1, and SF 2.

principal component (PC2)). In the fluorescence spectra, PC1 contributed 99.59% and PC2 0.38% of the 99.97% disparity.

Figure 11 displays the loadings or coefficient plot. It consists of the principal component coefficients (PCCs). The wavelengths are represented by the rows of the primary components, while the principal components are represented by the columns.

The loadings plot displays the range of 450–800 nm where the variations and similarities in the fluorescence spectra may be seen. It also shows how the four crude oil samples were grouped in a certain way. This map aids in illuminating the observations' trends. The plot indicates that the range of 470–570 nm is where the significant variances in the fluorescence spectra occur. The fluorescence spectra are clustered in PC space according to the two important PCs that were chosen, and the variance for the whole set of data is displayed in the Score plot in **Figure 12**.

Since sample JF was isolated from samples TF and SF along PC1, sample JF's crude oil may be heavier than samples TF and SF's. The variations in the fluorescence spectra of the JF, TF, SF 1, and SF 2 samples seen in the Score plot are explained by the negative loadings for PC2, which display a trough. It is consequently possible to report that the composition of crude oil varied from oil field to oil field as a result of the various geological sources from which the samples were collected. The PCA has demonstrated that the LIF of the crude oils from these four distinct wells can be distinguished using just two PCs.

19. Conclusion

Energy-dispersive X-ray fluorescence (EDXRF) and Laser-Induced Fluorescence (LIF) were employed. The samples' LIF spectra revealed five prominent peaks in the

510–750 nm spectral range, which is indicative of light, sweet crude oils with low specific gravities, often known as high API gravities. Hidden fluorescence peaks in the fluorescence spectrum were extracted by deconvolving the normalized LIF spectra of the samples. JF, TF, SF 1, and SF 2 were found to belong to five main fluorescence spectral groupings from the four different geological zones from which they were obtained. These groupings had distinct maximum peak positions at (510.00 $\pm$ 1.41) nm, (560.00 $\pm$ 2.83) nm, (612.50 $\pm$ 5.74) nm, (666.50 $\pm$ 13.87) nm, and (723.25 $\pm$ 19.41) nm. The varying concentration levels of the various fluorophores contained in the crude oil samples are reflected in these prominent fluorescence bands.

The collected LIF spectra were analyzed using rincipal Component Analysis (PCA) to determine the differences and similarities. Based on the LIF spectra, this unsupervised pattern recognition approach (PCA) classified JF, TF, SF 1, and SF 2 according to their geological positions and found cluster trends in the spectral data. As a result, PCA demonstrated that every crude oil sample came from a distinct geological zone.

Utilizing the energy-dispersive X-ray fluorescence method, the crude oil samples' sulfur content was determined and measured. It was possible to describe the samples according to the amounts of sulfur in them. All of the crude oil samples examined in this paper can be categorized as *sweet crude*, even though the data indicated that the Tweneboa Enyenra Ntomme (TEN) Oil Field had the greatest sulfur concentration.

Petroleum products are anticipated to differ because crude oil comes from a variety of geological sources. Four distinct oil fields provided four samples of crude oil. Studies have been conducted on Ghana's coastal arches' Jubilee oil fields (JF), TEN oil fields (TF), and Saltpond oil fields' Wells 2 (SF 1) and 4 (SF 2). Energy-dispersive X-ray fluorescence (EDXRF) and Laser-Induced Fluorescence (LIF) were the two spectroscopic methods employed.

This study sought to review how LIF and EDXRF can be employed as fast and straightforward spectrochemical techniques to characterize crude oils and refined fuel products, which confirmed that products of crude oil, such as petroleum fuels, aviation fuels, and liquefied petroleum gas, may also vary in physicochemical composition due to their different geological sources and thus require unique spectrochemical tools such as LIF and EDXRF due to their sensitivity, low-cost, and selectivity advantages over other methods.

Author details

Calvin Kwesi Gafrey*, Robert Wilson and Samuel Tetteh-Abaku
Department of Electronic Materials and Devices Engineering, Soonchunhyang University, Asan-si, South Korea

*Address all correspondence to: calvin.gafrey@stu.ucc.edu.gh

IntechOpen

© 2024 The Author(s). Licensee IntechOpen. This chapter is distributed under the terms of the Creative Commons Attribution License (http://creativecommons.org/licenses/by/4.0), which permits unrestricted use, distribution, and reproduction in any medium, provided the original work is properly cited.

References

[1] Pedersen M. Propagation of Shocks to Food and Energy Prices: An International Comparison. Charles University, Prague: Central Bank of Chile; 2011

[2] Evdokimov IN, Losev AP. Features of analysis of associative hydrocarbon media. Applicability of refractometric methods. Chemistry and Technology of Fuels and Oils. 2007;**43**:140. DOI: 10.1007/s10553-007-0027-5

[3] Hook M. Global Energy Systems [licentiated thesis]. Uppsala: Department for Physics and Astronomy, Uppsala University; 2009. pp. 3-13

[4] Tamrakar PK. Biomass and Volume Tables with Species Description for Community Forest Management. Vol. 101. Kathmandu, Nepal: Ministry of Forest and Soil Conservation; 2000

[5] Tamrakar PK, Pitre KS. Electrochemical determination of lead in gasoline. Bulletin of Electrochemistry. 2001;**16**(12):537-538

[6] Appenteng MK, Golow AA, Carbov D, Nartey V, Kaka EA, Salifu M, et al. Physicochemical characteristics of the Jubilee crude oil. Elixir Applied Chemistry. 2013;**54**:12513-12517

[7] Adipah S. Introduction of petroleum hydrocarbon contaminants and its human effects. Journal of Environmental Science and Public Health. 2019;**3**:1-9

[8] El-Hussein A, Marzouk A. Characterization of petroleum crude oils using laser-induced fluorescence. Journal of Petroleum & Environmental Biotechnology. 2015;**6**:5

[9] Gafrey CK, Wilson R, Amoako G, Anderson B. Laser-induced fluorescence with multivariate analysis in characterizing crude oils from some selected oil fields in Ghana. Physical Science International Journal. 2021;**25**: 8-20. DOI: 10.9734/psij/2021/v25i630262

[10] Wilson R, Gafrey CK, Amoako G, Anderson B. Metallic composition analysis of crude petroleum from some oil fields in Ghana. Physical Science International Journal. 2021;**25**(5):19-29. DOI: 10.9734/psij/2021/v25i530256

[11] UNEP/PCFV. Worldwide Sulphur Levels in Diesel Fuel. Nairobi, Kenya: UNEP/PCFV; 2009. Available from: http://www.kosmosenergy.com/ghana.html

[12] Qasim M, Ansari TM, Hussain M. Physicochemical characteristics of Pakistani used engine oils. Journal of Petroleum Technology and Alternative Fuels. 2016;**7**:13-17. DOI: 10.5897/JPTAF.2015.0121

[13] Gafrey CK. Comparative Studies of Crude Oils from Some Selected Oil Fields in Ghana Using Spectroscopic Techniques [thesis]. Ghana: University of Cape Coast; 2021. Available from: https://scholar.google.com/citations?view_op=view_citation&hl=en&user=jhqxYdsAAAAJ&authuser=1&citation_for_view=jhqxYdsAAAAJ:d1gkVwhDpl0C

[14] Abdalla SS. Determination of Hydrogen Content of Petroleum Products from Tema Oil Refinery Using Neutron Backscatter Technique [thesis]. Ghana: University of Ghana, Legon; 2015. Available from: https://ugspace.ug.edu.gh

[15] Frontasyeva MV. Neutron activation analysis in the life sciences. Physics of Particles and Nuclei. 2011;**42**:332-378

[16] Heftmann E. Fundamentals and applications of chromatography and related differential migration methods. Journal of Chromatography Library. 2004;**69**(Part B):519-1156

[17] Svanberg S. Atomic and Molecular Spectroscopy: Fundamental Aspects and Practical Applications. Vol. 6. Berlin, Heidelberg: Springer Science & Business Media; 2012

[18] Hollas JM. Modern Spectroscopy. 4th ed. Hoboken, New Jersey, USA: John Wiley & Sons; 2004. ISBN 0-470-84416-7

[19] Sauer M, Hofkens J, Enderlein J. Handbook of Fluorescence Spectroscopy and Imaging: From Ensemble to Single Molecules. 1st ed. Hoboken, New Jersey, USA: Wiley-VCH; 2012

[20] Lakowicz JR. Principles of Fluorescence Spectroscopy. New York City, USA: Springer Science & Business Media; 2013

[21] Lakowicz JR. Solvent and environmental effects. In: Principles of Fluorescence Spectroscopy. Boston, MA: Springer; 2006. DOI: 10.1007/978-0-387-46312-4_6

[22] Pavia LD, Lampman GM, Kritz GS, Vyvyan AJ. Introduction to Spectroscopy. Edition 4, illustrated ed. Bellingham, Washington, USA: Cengage Learning; 2008

[23] Balshaw-Biddle K. Subsurface Monitoring Using Laser Fluorescence. Boca, Raton: Lewis Publishers; 2000

[24] Baumann T, Haaszio S, Niessner R. Applications of a laser-induced fluorescence spectroscopy sensor in aquatic systems. Water Research. 2000; **34**:1318-1326

[25] Schultze R, Lemle M, Löhmannsröben HG. Laser-Induced Fluorescence (LIF) Spectroscopy for the *in situ* Analysis of Petroleum Product Contaminated Soils. Heidelberg: Springer; 2004. pp. 79-98

[26] Lömannsröben HG, Roch T, Schultze RH. Laser-induced fluorescence spectroscopy for *in situ* analysis of petroleum products and biological oils in soils. Journal of Polycyclic Aromatic Compounds. 1999;**13**:516-587. DOI: 10.1080/10406639908020561

[27] Praider B, Largeau C, Derenne S, Martinez L, Betrand P. Chemical basis of fluorescence alteration of crude oils and kerogens. Microfluorimetry of oils and their fractions. Organic Geochemistry. 1990;**16**:451-460

[28] Bujewski G, Rutherford B. The Rapid Optical Screening Tool (ROST) Laser-Induced Fluorescence (LIF) System for Screening Petroleum Hydrocarbons in Subsurface Soils. Las Vegas: Innovative Technology Verification Project Report; 1997

[29] Environmental Protection Agency [EPA], Research& Development. 1997. Available from: https://www.epa.gov/superfund/health/contaminants/radiations/pdfs/umtrcagu.pdf.

[30] Takumi S, Lukman S, Aoyagi N, Takaumi K, Nagasaki S. Speciation of Eu^{3+} bound to humic substances by time-resolved laser fluorescence spectroscopy (TRLFS) and parallel factor analysis (PARAFAC). Geochimica et Cosmochimica Acta. 2012;**88**:199-215

[31] Collins RN, Saito T, Aoyagi N, Payne TE, Kimura T, Waite TD. Applications of time-resolved laser fluorescence spectroscopy to the environmental biogeochemistry of

actinides. Journal of Environmental Quality. 2011;**40**:731-741

[32] Schade W, Bublitz J, Helbig V, Nick KP. In: Werner C, Klein V, Weber K, editors. Time-Resolved Laser-Induced Fluorescence Spectroscopy for Diagnostics of Oil-Pollution in Water. Berlin, Heidelberg: Springer; 1992. DOI: 10.1007/978-3-642-50980-3_8

[33] Hercules DM, Joyner DJ. Chemical bonding and electronic structure of B_2O_3, H_3BO_3 and BN: An ESCA, auger, SIMS, and SXS study. The Journal of Chemical Physics. 1979;**72**:1095

[34] Daućik P, Zidek Z, Kalab P. Determination of sulphur content in fuels. Chemical Papers. 1998; **52**(5):667-670

[35] Demtröder W. Laser Spectroscopy: Basic Concepts and Instrumentation. New York City, USA: Springer Science & Business Media; 2013

[36] Martinez WL, Martinez AR. Exploratory Data Analysis with MATLAB. In: Computer Science and Data Analysis Series. New York City: Chapman & Hall/CRC; 2005

Section 4

LPG and BioLPG: Current Status and Future Prospects

Chapter 10

Perspectives of a Sustainable Future Based on LPG and Renewable BioLPG as Fuel Suppliers: New Developments and Existing Barriers

Lina Montuori and Manuel Alcázar-Ortega

Abstract

This chapter provides an overview on the outstanding production, distribution and storage technologies that will likely make the liquefied petroleum gas (LPG) and renewable BioLPG expand their utilization as a cleaner alternative to traditional fossil fuels. The growth of the Global LPG Market size will be presented, with a focus on US and the Europe, the Middle East and Africa (EMEA) regions. Moreover, the increment of worldwide policies aimed at reducing greenhouse gas emissions (GHGs) will be investigated. The growth of the LPG market is primarily propelled by the rising demand for a cost-effective and environmentally sustainable substitute of conventional fossil fuels. Concerning the residential sector, the demand for LPG is due to such end-uses as cooking, water heating and other household purposes. Regarding transportation, LPG's popularity as an alternative automotive fuel (auto-gas) has also contributed significantly to its widespread adoption in this sector. On the other hand, the utilization of BioLPG is also going up since it is chemically indistinct from LPG and so, it can be 'dropped-in' to existing LPG boilers and appliances. Finally, the existing barriers and the new developments on this topic will be investigated to understand why LPG and BioLPG could represent a great investment for the future.

Keywords: liquefied gas petroleum, BioLPG, barriers, sustainability, renewable energy, mobility

1. Introduction

In a world grappling with the challenges of global warming and climate change, liquefied petroleum gas (LPG) is gaining popularity as a clean and sustainable alternative source to carbon fossil fuels. Primarily composed of propane and butane [1], LPG offers a cleaner burning alternative to traditional fossil fuels, such as coal and oil [2]. It is due to the fact that, when ignited, LPG emits significantly lower levels of harmful pollutants, including carbon dioxide (CO_2), nitrogen oxides (NOx) and

 IntechOpen

particulate matter (PM). Now, according to the World LPG Association (WLPGA) [3], LPG emits up to 15% less CO_2 than coal and 30% less than oil when used for heating and cooking purposes. This is because LPG produces less CO_2 emissions overall. Compared to gasoline, previous studies have demonstrated that LPG produces, respectively, 69.94%, 46.48%, and 47.89% lower hydrocarbon (HC), NOx and carbon monoxide CO emissions. The lower HC emissions have been demonstrated to be due to the higher flame's propagation speed and combustion temperature of LPG [4, 5].

Furthermore, the impact that LPG has on air pollution is far lower than natural gas when used in spark-ignited engines. Previous comparative studies showed as direct consequence of its considerably higher methane (CH_4) emissions that compressed natural gas generates a 10% and 4% higher brake-specific carbon dioxide equivalent ($bsCO_2eq$) at low engine loads than LPG. Moreover, LPG fuel consumption is lower when producing the same amount of energy, as it has a higher energy content compared to natural gas [6]. Indeed, the European Commission approved a long-term strategy for driving EU towards a net-zero GHG economy with the target of reducing emissions by 80–95% in 2050. As part of the European Green Deal, the Commission proposed on 4 March 2020 the first European Climate Law to enshrine the 2050 climate-neutrality target into law [7].

In summary: the worldwide awareness that reduction in greenhouse gas emissions plays a vital role in mitigating climate change and improving air quality has driven LPG to be considered a crucial tool in the fight against global warming. Therefore, LPG is a perfect candidate to be the clean alternative to petroleum in the medium term for the transition to sustainable fuels and transport [8].

1.1 BioLPG, the renewable and sustainable form of LPG

Some studies, such as Ref. [9], showed how switching fuels to LPG in the immediate term and, progressively, to BioLPG as a direct replacement fuel, can contribute to establish pathway towards decarbonization for industrial and commercial businesses. BioLPG is a renewable and sustainable form of LPG derived from organic waste materials, agricultural residues and other biomass sources with a carbon intensity of around 70–80%, which is lower than that for oil. It is also called renewable propane and biopropane, as it can reduce the carbon footprint and CO_2 emissions up to 80% [10].

BioLPG can be produced in several ways depending on the feedstock and processes used. In **Figure 1**, the main process currently in use, together with their technology readiness level (TRL), is presented.

As it can been seen from **Figure 1**, among all promising production processes, hydroprocessing of vegetable oils (HVO) is the most mature one and, currently, it is the main production route for biopropane, able to produce (via HVO and fats) significant volume of HVO BioLPG with a process performance in the range of 5–8% [11]. Therefore, there is still significant work ahead to implement and test novel technologies able to produce BioLPG at commercial scale and at affordable costs to end users [10].

The growing interest in BioLPG as renewable fuel can be attributed to the following factors: First, BioLPG is produced by renewable crops grown and waste feed stocks, which made it an ideal energy solution to meet the EU sustainable goals and enable the transition to a circular economy. On the other hand, BioLPG offers similar performance and efficiency as conventional LPG as both are chemically identical. Consequently, it is compatible with all LPG applications and it can be blended and used by all existing LPG

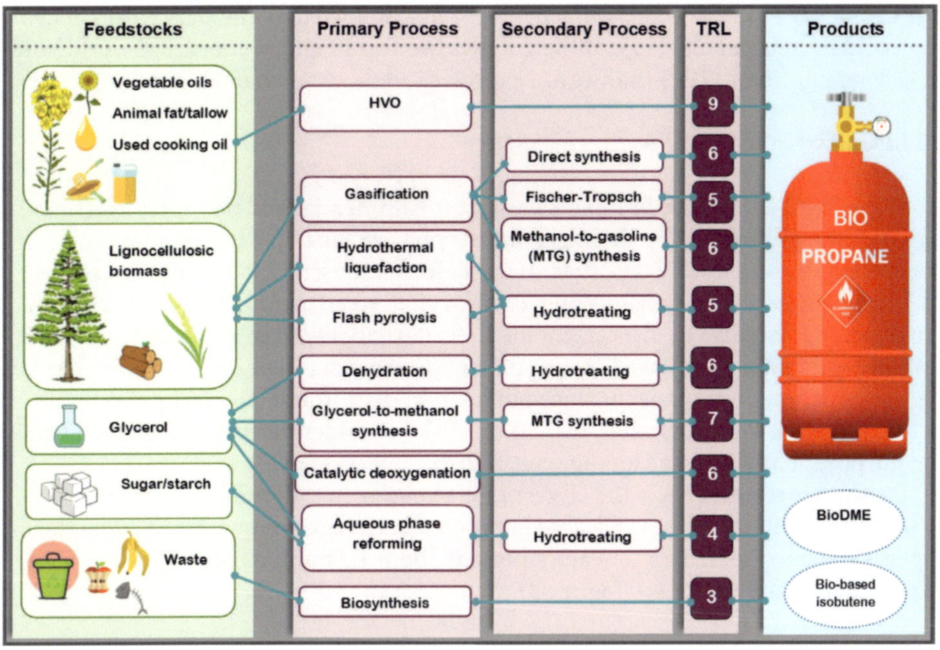

Figure 1.
Feedstocks and processes for biofuel production. Source: Liquid Gas UK [11].

equipment without significant modifications. Moreover, BioLPG offers the same benefits as conventional LPG in terms of high quality, low-carbon fuel with reduced emissions and improved air quality, but with the added advantage of being renewable and carbon neutral. Furthermore, the growing concerns about greenhouse gas emissions resulted worldwide in the implementation of supportive policies and incentives to encourage the adoption of cleaner fuels as in the UK, where Liquid Gas UK (LGUK) has proposed BioLPG as a potential solution to satisfy residual gas needs for off-grid heating to reduce carbon emissions [12]. These initiatives, coupled with the increasing availability and affordability of BioLPG, have contributed to the rise of its use among industries, households and the transportation sector [13].

2. Current trends of the LPG market in US and EMEA regions

Liquefied petroleum gas, together with its eco-friendly sibling, BioLPG, is playing a crucial role in reducing greenhouse emissions. To understand what delves into the reasons behind the growing popularity of LPG and BioLPG as a competitive clean fuel, it is important to be aware of the market trends.

Especially in developing countries, LPG has the prevalent usage as household fuel [14]. Such countries as India, Indonesia, Morocco or some South American nations still rely on LPG on a daily basis, while others like some countries in Africa aim to increase household LPG use up to 80% by 2030 [9].

The recent trend in LPG consumption worldwide shows that, on average, it is equal to 54.28 thousand barrels per day (picture from 2021) based on the data belonging to 185 countries. The country with the major consumption of LPG in 2021 was China with 2258.97 thousand barrels per day, followed by the USA with 1374.98 thousand barrels per day [15]. **Table 1** shows the Top 10 global countries' worldwide LPG consumption in 2021 (values are expressed in thousand barrels per day).

2.1 LPG market in the USA

In the USA, LPG is predominantly used for commercial and domestic use, heating or cooking, followed by chemical industry where the LPG's largest market is petrochemical feedstock. Moreover, in agriculture industry, LPG is largely used for fueling machinery, livestock-rearing, crop-drying, weed and animal waste disposal. Furthermore, it is also used as an engine fuel alternatively to gasoline and as a stand-by fuel for facilities that have interruptible natural gas service contracts [16]. The U.S. field production of propane increased in the past decade and it is forecasted to rise by another 5% in 2024. Altogether, the use of LPG for residential and commercial heating purposes has witnessed an increase of 20%, as shown in **Figure 2**.

Furthermore, the USA is the largest LPG exporter in the world. LPG exports reached 14.5 million tonnes of seaborne volume in the second quarter of 2023, the highest quarterly total ever to reach a peak of about 1.7 million barrels per day (b/d) exported in March 2023 (see **Figure 3**).

2.2 LPG market in Europe

In Europe, the versatility of LPG as fuel that can be easily liquefied and stored in multiple ways allowed its use in different sectors, such as space and water heating, cooking, lighting, power generation, industrial processing, and heating and automotive fueling.

In 2023, Europe's LPG demand stood at 46.79 million tonnes and it is forecasted to reach the 77.68 million tonnes roughly in 2032 [17]. The aviation industry and its use for domestic cooking are predominant in European regions but the rising consumption of this fuel in the automotive transportation is driving the growth of the LPG market. In fact, auto-gas consumption (as it is called LPG fueled in cars) has

Global rank	Countries	LPG consumption, 2021 (thousand barrels per day)
1	China	2258.97
2	USA	1374.98
3	India	886.55
4	Russia	802.34
5	Saudi Arabia	494.83
6	Japan	404.53
7	South Korea	299.92
8	Mexico	278.86
9	Indonesia	271.86
10	Thailand	194.80

Table 1.
Top 10 global countries' worldwide LPG consumption in 2021 [15].

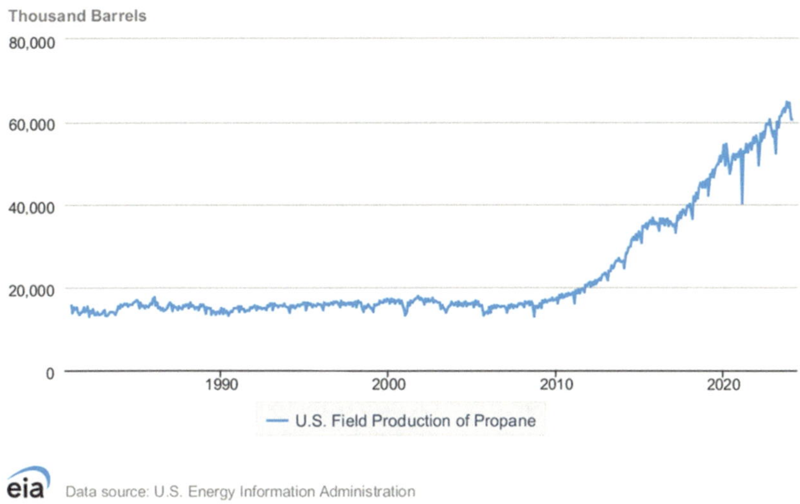

Figure 2.
U.S. field production of propane in thousand barrels. Source: EIA.

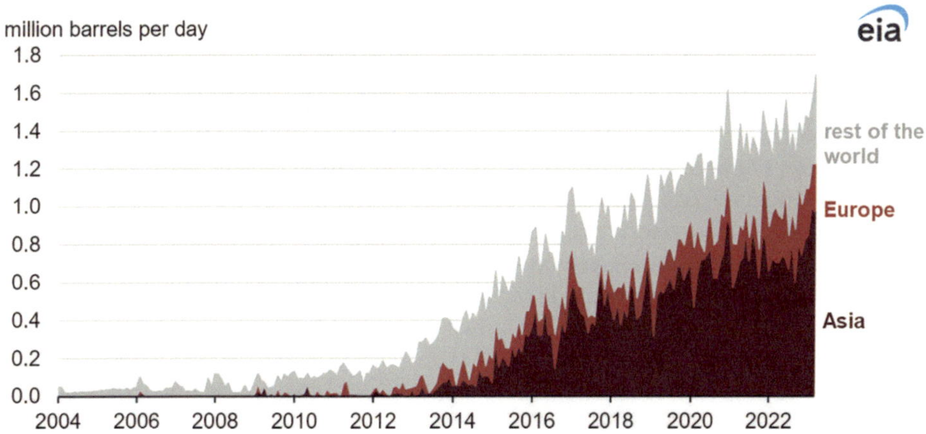

Figure 3.
Monthly U.S. propane exports by destination region—January 2004-March 2023. Source: Environmental Impact Assessment (EIA).

increased more than 10% in the past 5 years, with BioLPG accounting for a significant portion of this growth [18].

Vehicle emissions of millions of EU citizens (including cars and vans) have been accounted to be, respectively, for around 12% and 2.5% of the total EU emissions of CO_2, respectively. Therefore, EU governments are promoting initiatives to bolster the use of LPG as retrofit fuel for internal combustion engine vehicles, for large vehicles and fleet management solutions [19].

Liquefied petroleum gas, together with BioLPG, can represent a green alternative to fossil fuels to meet the EU sustainable targets, especially in urban areas. The 2003 European Emission Testing Program (EETP) demonstrated that LPG-fueled vehicles are 120–180% better than petrol and 2000% better than diesel in terms of NOx. Moreover, it is 99% better than diesel in terms of particle emissions what is the clear advantage of its use [20].

Recent issues, like COVID-19 and Ukraine war, highlighted the need for building reliable coordinating measures to secure energy supply. The Directive 2009/119/EC requires EU Member States to maintain minimum stocks of crude oil and/or petroleum products. Likewise, the membership obligations of the International Energy Agency (IEA) include maintaining oil reserves equivalent to 90 days of net imports [21]. Moreover, the Russian conflict pinpoints the potential of LPG as natural gas substitute in case of disruption due to their physical similarity and compatibility with the existing infrastructure. In this framework, EU self-sufficiency can be promoted by adopting LPG for three main reasons: due to its large availability, the overall balance in EU consumption and production and the geographical diversifications of the imports [20].

As it is well known, LPG can be derived from a plurality of sources such as crude oil and natural gas, and its possible production from biomass (BioLPG), which has lately attracted government attention worldwide. Commonly transported as liquid in different ways, among its advantages there is the fact that it does not rely completely on a fixed reticulated and pipeline infrastructure as it happens with natural gas. Rather, it can be transported by using a combination of ways, like ships, road tankers and tanker rail cars. These suitable characteristics, together with extensive availability of LPG and its flexible and decentralized distribution network (as shown in **Figure 4**), enable its use in rural and isolated areas that are not served by other energy fuels.

The LPG industry allows import, storage and pressurized transport of LPG everywhere and it is especially significant in areas that are not served by the natural gas distribution infrastructure. Accordingly, **Figure 4** shows the LPG supply chain from the production site at refinery or gas receiving terminals to the end user. LPG can be imported by pipeline, rail and road or can be containerized. The bulk of storage can be refrigerated at atmospheric pressure with a temperature of −43°C or pressurized at ambient temperature with a pressure from 4.0 to 13.0 bar. Domestic distribution of LPG also takes place by road transport, rail, pipelines or ISO (International Organization for Standardization) container on road or by sea.

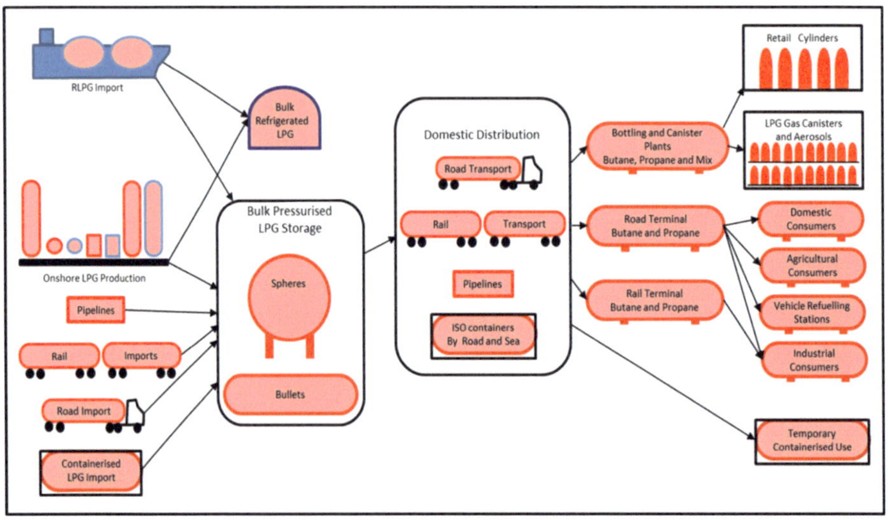

Figure 4.
Map of the LPG industry and distribution. Source: European Commission Joint Research Centre (JRC) [22].

Regarding the LPG consumption, European countries (based on 27 countries) consumed 29.71 thousand barrels per day in 2021, as shown in **Figure 5**. The highest value was for Germany, with 116.56 thousand barrels per day, while the lowest value was for Luxembourg with 0.33 thousand barrels per day. Only a few European countries have included LPG in their energy mix. Germany, France and Italy have strongly embraced LPG usage. The reason why Germany and France decided to rely on LPG, mainly for the automotive sector, which resides on their market trend that is less affected by prices of inflation (like gasoline) and, consequently, LPG results as a convenient, alternative fuel [23]. Moreover, the tough rise of gas price has stressed this tendency and prompted many refineries to switch to LPG as a cheaper alternative. In Italy, LPG usage has grown in the last years due to a strategic decision of adopting BioLPG as a renewable fuel before shifting to electric car or heat pump. Nowadays, BioLPG consumption has reached the 50% in the automotive sector, 25% for heating and cooking and the remaining by the industrial process. Italy is trying to increase LPG production, especially BioLPG, from organic sources, which is expected to reach the 5 million metric tonnes/year by 2030 [3].

The EU production is slightly lower than the demand, needing a 6% of LPG net import from abroad (**Figure 6**) but LPG supply is expected to exceed by far the projected demand by 2030 [24].

In Europe, the existence of a highly fluid international trading market guarantees that LPG imports are geographically diversified. **Figure 7** shows the geographical diversification of the European LPG imports, as well as the export terminal locations. In the first 10 months of 2022, the European Union significantly increased its seaborne imports of LPG from the United States, reaching 6.4 million metric tonnes between January and October 2022, which represents almost twice the same period in 2021 [24].

The current growth of the LPG market is mainly propelled by the rising demand of a cost-effective and environmentally sustainable substitute for conventional fossil

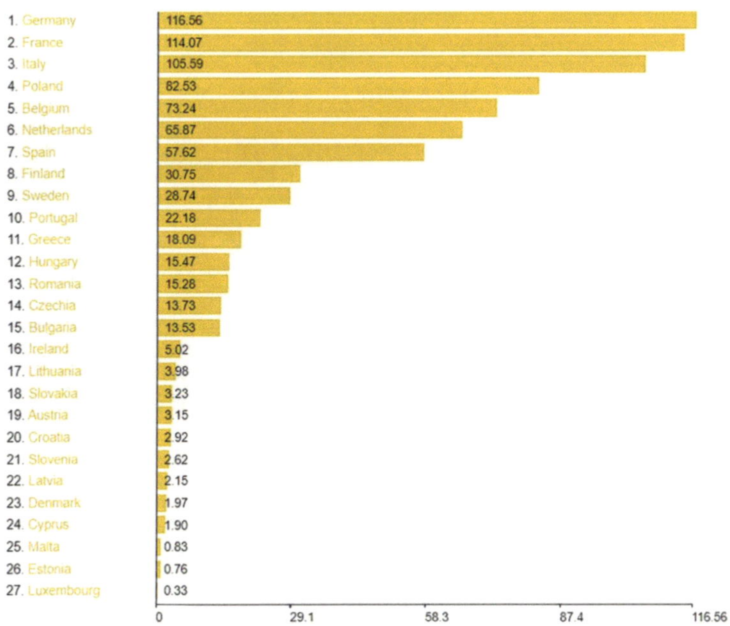

Figure 5.
Liquefied petroleum gas consumption (2021), in thousand barrels per day. Source: Environmental Impact Assessment (EIA).

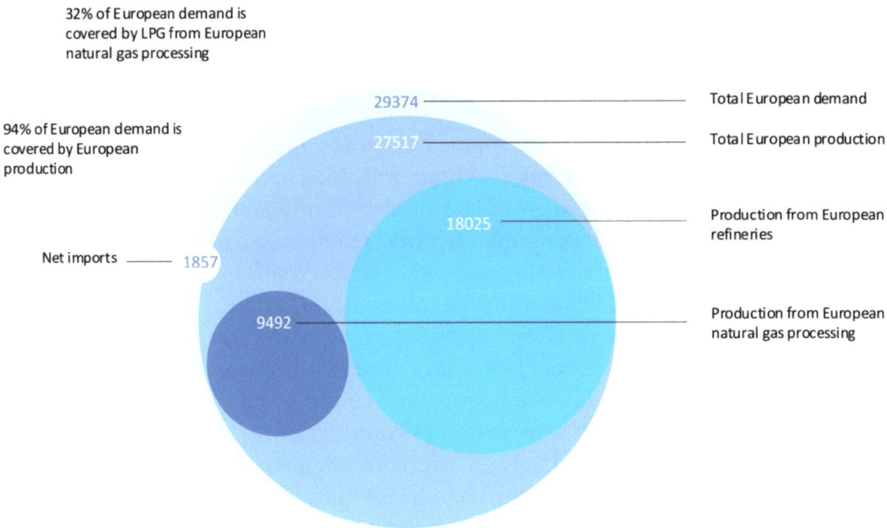

Figure 6.
Supply vs. demand European LPG (thousands of tonnes). Source: World LPG Association (WLPGA).

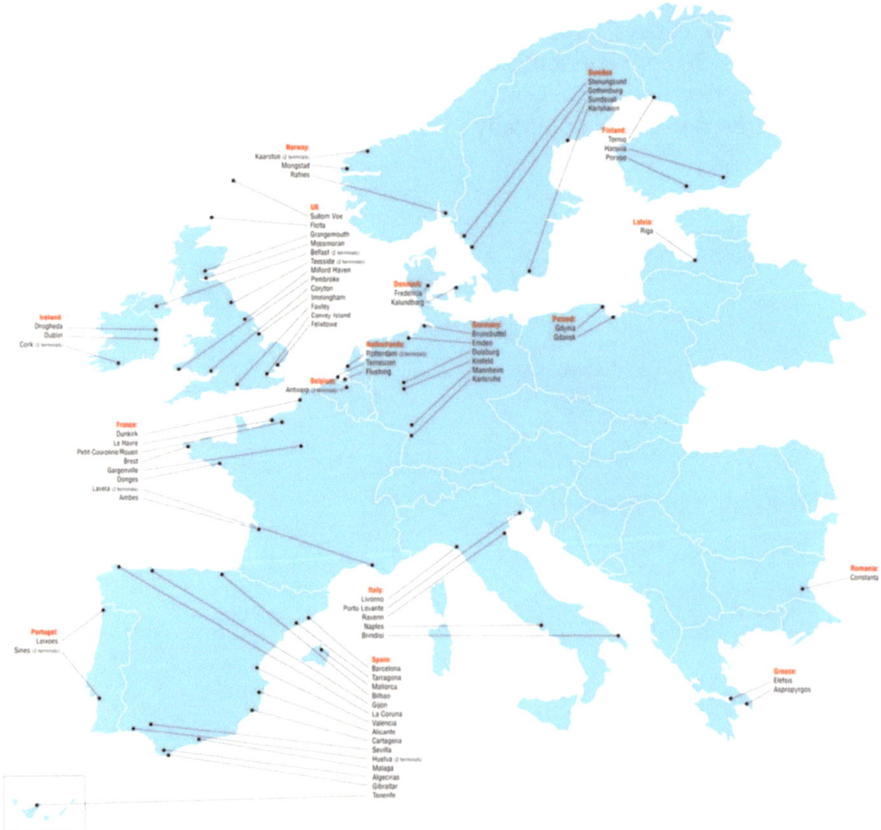

Figure 7.
Key European LPG import and export terminal locations. Source: World LPG Association (WLPGA).

fuels. In this framework, BioLPG is gaining popularity too since it is chemically indistinct from LPG and so, it can be 'dropped-in' to existing LPG boilers and appliances. In spite of that, the BioLPG market is still limited and, indeed, if compared with LPG global production, BioLPG production is still less than 0.1% [25]. However, it is expected to grow over 2.2 metric tonnes per year (Mt/yr) by 2030 [9].

3. The role of the LPG and BioLPG towards more sustainable mobility in Europe

Automotive LPG, also known as Autogas, constitutes together with BioLPG a dynamic duo widely used in the EU as alternative to fossil fuel. Benefits of using Autogas are related to its large availability, safety, lower emissions and the lower cost they entail if compared with gasoline and diesel (**Figure 8**). Studies have demonstrated that Autogas-fueled vehicles emit over 20% less carbon monoxide, 10% less CO_2 and more than 40% less nitrogen oxides than gasoline engines [26]. Moreover, minimum adaptations to the existing infrastructure are required to use LPG, with minimum investment in infrastructure [27].

Figure 9 provides an overview of the average Autogas price, showing that it is more convenient, from the economic point of view, than fossil fuels (petrol and diesel). As it can be seen, Autogas is an appealing solution towards the green EU pathways as it is an economic alternative. The large availability of this fuel and the surplus of its supply, able to afford the ensuing growth in the demand together with favorable tax rate established by the EU Directive, are the reasons why LPG is the most promising and economic fuel alternative for the automotive sector's decarbonization [28].

Nowadays, as shown in **Figure 10**, there are about 8.5 million registered vehicles (new and retrofit passengers' cars) fueled with LPG in the EU and over 30,700 Autogas filling stations [29]. This increment of about 16% in Autogas vehicles, compared to the previous year, is due to the ongoing energy transition that has identified

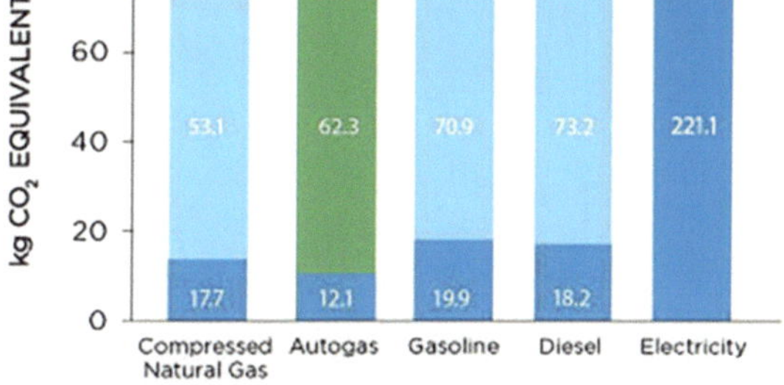

Figure 8.
Total carbon emissions for various fuels. Source: Alliance AutoGas.

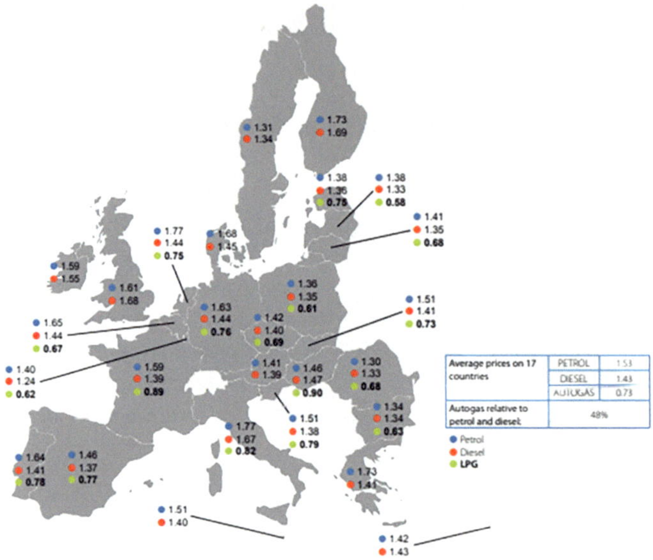

Figure 9.
Overview of the average pump price in a euro per liter (€/l) (value-added tax (VAT) and exercise duty included) in a selection of EU countries (2013). Source: Oil Bulletin, Directorate-General for Mobility and Transport (DG MOVE), European Commission.

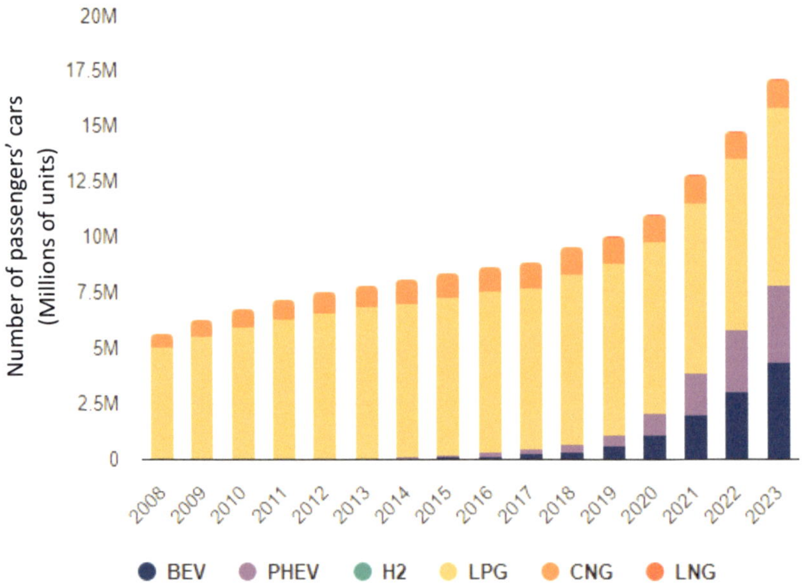

Figure 10.
The total number of alternative fueled passengers' cars in the EU. Source: Alternative Fuels Observatory [29]. Considered types of cars are: battery electric vehicles (BEVs); plug-in hybrid electric vehicles (PHEVs); vehicles fueled by hydrogen (H_2); liquefied petroleum gas (LPG); compressed natural gas (CNG); and liquefied natural gas (LNG).

the automotive sector as responsible for the almost 20% of Europe's total greenhouse gas emissions (GHGs) [30].

Autogas market is well developed worldwide. LPG demand reached about 46.79 million tonnes in 2023, although it is forecast to grow sensibly in the next future reaching 69.89 million tonnes by 2030 [31]. According to the countries where the consumption of Autogas is more significant, just five countries (Turkey, Russia, South Korea, Poland and Ukraine) together accounted for almost a half of the global Autogas consumption in 2021, as shown in **Figure 11**.

Figure 11 also shows a significant increment of Autogas consumption from 2010 since, in that year, a re-categorization of LPG demand assigned before to the residential sector was produced.

Regarding the USA, Autogas has been used in transportation since 1912 thanks to the California Energy Commission, which encouraged Autogas vehicle conversions and refueling stations. After that, Autogas was declared an alternative fuel under the Energy Policy Act of 1992 [32]. Today in the USA, there are about 155,000 road vehicles powered by Autogas and about 2500 public filling stations with available fuel for use in vehicle distributed on all US territories (**Figure 12**).

Together with the LPG, opportunities for feeding the existing vehicles with BioLPG are also under evaluation in the USA. Renewable propane (BioLPG) obtained using renewable or bio-based feedstock is nowadays used as drop-in replacement fuel for conventional propane [34]. The LPG's low price, the reduced maintenance cost and the minimum environmental impact, together with the low level of noise pollution, made Autogas engines attractive for urban transportation. In 2014, in the USA, it was estimated that more than 7000 busses for pupil transportation were reconverted to Autogas propane engines [35].

The spread out of federal and state regulations to stimulate Autogas applications has brought about the wide spread of vehicle conversions and the increase in the refueling stations, but the American Autogas market is still small when compared to other automotive fossil fuel markets [36]. The chart in **Figure 13** shows

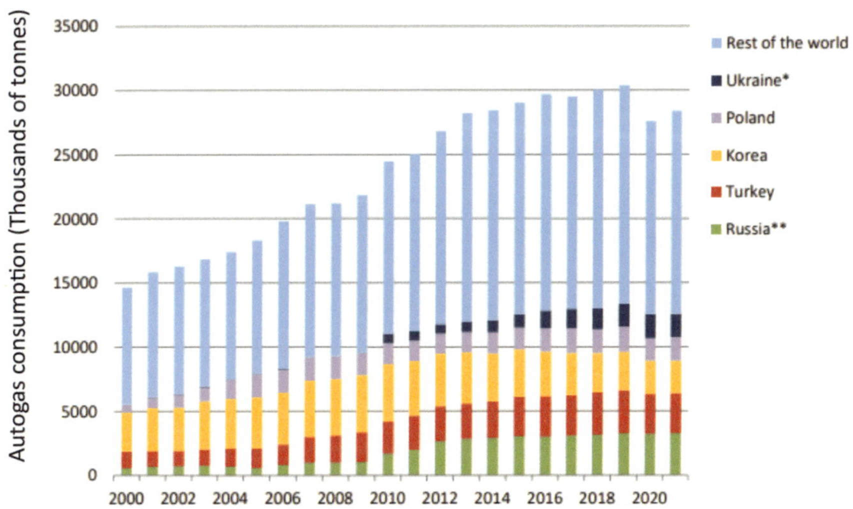

Figure 11.
World Autogas consumption in 2000–2021. Source: World LPG Association (WLPGA)/Argus (2022); International Energy Agency (IEA) databases.

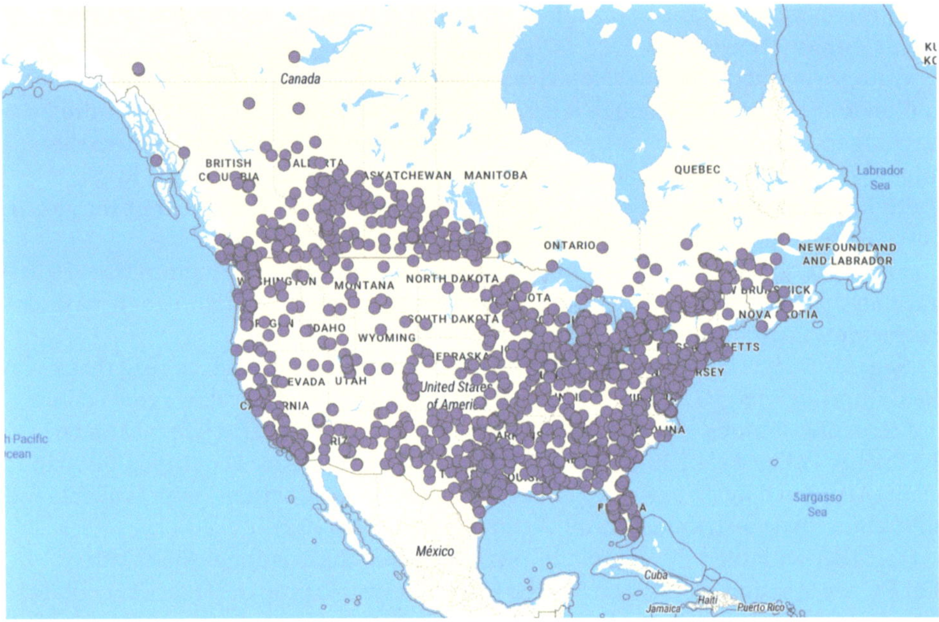

Figure 12.
Propane filling stations' location by the US state. Source: Alternative Fuels Data Center [33].

the trend in the USA vehicle registration from 2016 up to 2020. Hydrogen fuel cell light-duty vehicles show the greatest growth, reaching the maximum peak in 2019, while the annual change in propane is more volatile as propane vehicles are mostly used for school bus fleets. Electric vehicle registrations had seen a steady growth until 2020, but they grew more than any other light-duty vehicle type between 2021 and 2022 [37].

The US federal and state regulations and incentives have had a significant impact on the wide spread of LPG-driven vehicles, as evidenced in **Figure 14**. Tax credits, low-interest loan programmes and tax exemptions adopted by different States reduced the LPG cost of conversion and offered more incremental savings to end

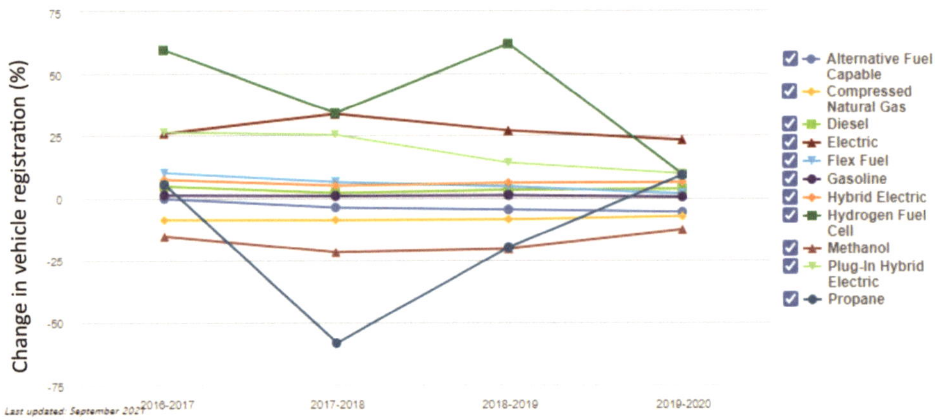

Figure 13.
Change in US vehicle registration counts. Source: Alternative Fuels Data Center.

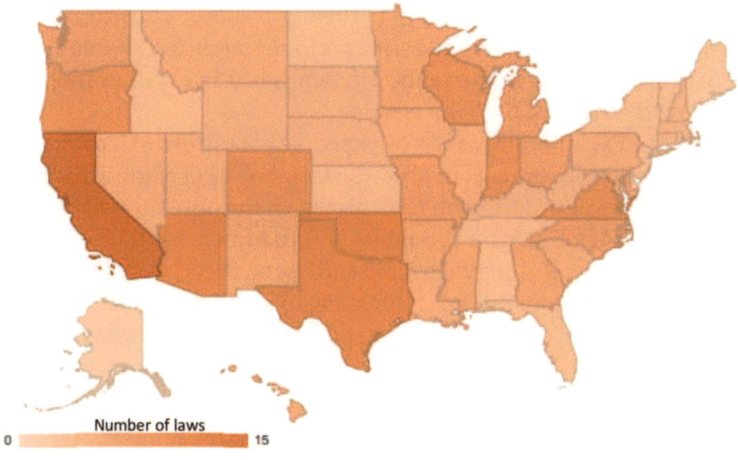

Figure 14.
Propane laws and incentives by state. Source: Alternative Fuels Data Center.

users [38]. The introduction of lower fuel tax rates on LPG has resulted in a lower maintenance cost for LPG vehicles than petrol or diesel ones. Moreover, the introduction of the National Energy Policy Act 1992 obliged US citizens to shift to more sustainable fleets fueled by alternative fuels (among them, LPG). In terms of safety, the design and construction of LPG tanks have been regulated by US government agencies and handling protocols have been approved for personnel training [39].

4. Existing barriers and new developments in LPG and BioLPG as fuel for a sustainable future

The adoption of LPG and BioLPG as clean fuel alternatives in Europe and the USA faces several challenges and barriers that need to be addressed.

The growing global awareness and concern about global warming and greenhouse gas emissions have led governments worldwide to implement regulation and subsidies to start a sustainable transition towards alternative and green fuels and, among them, LPG and renewable LPG. Despite that, additional incentives should be established to increase the use of alternative fuels for vehicles and infrastructure. Governments should work closely with fuel providers and fleet managers to offer incentives able to make drivers switch to alternative fuels and increase LPG applications.

Autogas, as alternative fuel for light-duty vehicles in the EU, is facing relevant challenges related to current regulation, which bans the use of internal combustion engines after 2035. Germany weakened the law by introducing an exception for synthetic e-fuels; Italy, Romania and Bulgaria decided to remain neutral and just Poland voted against that legislation. In this context, the European Commission is developing new legal rules for allowing just the sale of cars powered with e-fuels after 2035. Unfortunately, the role of Autogas in this scenario is not clear, as CO_2 neutral fuels are not explicitly identified in the legislation [30].

Additional studies should be carried out to evaluate the emission rate of the existing vehicles fueled with LPG and, especially, BioLPG to demonstrate their low environmental impact and to make grow people's awareness about the potential benefits in quality of air when using LPG. There are studies ongoing with the purpose of evaluating the benefits of LPG and BioLPG fuels in terms of emissions of secondary organic aerosols (SOA). Moreover, investments in research and development should be done by manufacturers to improve the efficiency of LPG system and reinforce refueling LPG infrastructure for vehicles [34].

Forecast about LPG automotive sector expects a steady growth of the automotive market in the coming years. However, the trend of LPG engine markets is still affected by the impact of inflation on LPG prices. In Europe, the current price policy undertaken by each Member State should be able to mitigate inflation effects on the LPG price, as it is still the driving factor for end users' fuel choice [40].

Among the new developments, it is interesting to highlight the latest advancements in fuel cell production as eco-friendly power source that extracts hydrogen from LPG. Although hydrogen is commonly produced from natural gas, steam reforming is a technology that can be used to produce hydrogen from other fuels like propane [41].

Manufacturers are exploring new fuel cell technologies and new ways of producing hydrogen by using renewable energies or alternative fuels. Indeed, propane can be a valid alternative due to its low emissions when compared with other existing fuels [42]. Accordingly, different studies have demonstrated the possibility of producing high concentrations of hydrogen from LPG reforming. Actually, during the final state of propane cleavage of the steam reforming, the reaction of oxidative reforming of LPG resulted in a lower coke formation, a higher production of H_2 with a yield value that can vary from 77.5% to 92.2%. On the other side, the most suitable conditions for the process required the use of an excess of water (H_2O/LPG = 7.0) and intermediate temperatures (973 K) [43–46].

Among the studies focused on the use of propane in different fuel cells, the special suitability of propane for direct fuel cells (DPFCs) has been identified. The two main reasons for propane's choice as an alternative fuel rely on the fact that infrastructures for direct propane fuel cells are already available in rural areas as propane is still predominantly used there for heating. Moreover, in rural areas, the power price produced by DPFCs is still competitive compared to the one of grid-connected utilities. Furthermore, propane is available everywhere and it can count on a reliable and robust infrastructure, while hydrogen distribution infrastructure is not ready yet. Additionally, the lower cost of propane as fuel for hydrogen production can positively affect the cost of manufacturing hydrogen. Ultimately, hydrogen storage is very expensive while propane can be easily stored and transported [47, 48].

Regarding BioLPG, its use has been tested in automotive applications with positive results and it is widely supported by regulation worldwide. On the contrary, for power generation applications, it is still required to carry out the implementation of subsidiary policies able to incentivize and boost power generation from renewable propane. BioLPG is characterized for lower emissions than conventional diesel; it is practically neutral and it allows a reduction of 3 tonnes of CO_2 for each tonne of BioLPG consumed. In this scenario, the framing of a supportive incentives regulation for LPG power applications can actively contribute to the energy transition and accelerate decarbonization [49].

5. Conclusions

This chapter highlights the advantages of LPG as a sustainable alternative to carbon fossil fuels due, among other benefits, to the lower emissions of CO_2 (15% less than coal and 30% less than oil) and other pollutants. This fact may help reach international commitments assumed by different countries for environmental reasons, such as the target of reducing emissions of the European Commission by 2050. Moreover, the so-called BioLPG (that is a renewable and sustainable form of LPG derived from organic waste and biomass sources) can help to reduce the carbon footprint up to 80%, considering the life cycle of the used materials.

Both LPG and BioLPG are mainly used for residential applications (mainly cooking and space heating) and for mobility issues to fuel cars. The automotive LPG, also known as Autogas, presents several benefits such as large availability, safety, lower cost and lower emissions. In fact, Autogas-fueled vehicles emit over 20% less CO, 10% less CO_2 and 40% less NOx than gasoline engines.

In spite of LPG's benefits, there are still barriers that prevent the massive utilization and further development of this fuel in terms of sustainability. Many of them are regulatory since the role of LPG in the framework of future clean energy supply has not been clearly stated, with a lack of regulation existing in that direction to promote the utilization of this fuel in the short term. In the automotive area, one of the most promising future applications of LPG is the obtaining of hydrogen to be used in combustion engines or fuel cells. In summary: Properties of LPG and its variants (BioLPG and Autogas) make a good solution to accelerate decarbonization and contribute to the energy transition for a future cleaner and sustainable energy production.

Author details

Lina Montuori* and Manuel Alcázar-Ortega
Universitat Politècnica de València, Institute for Energy Engineering, Valencia, Spain

*Address all correspondence to: lmontuori@iie.upv.es

IntechOpen

© 2024 The Author(s). Licensee IntechOpen. This chapter is distributed under the terms of the Creative Commons Attribution License (http://creativecommons.org/licenses/by/4.0), which permits unrestricted use, distribution, and reproduction in any medium, provided the original work is properly cited.

References

[1] Finlayson-Pitts BJ, Pitts JJ Jr. Applications of atmospheric chemistry: Air pollution control strategies and risk assessments for tropospheric ozone and associated photochemical oxidants, acids, particles, and hazardous air pollutants. In: Chemistry of the Upper and Lower Atmosphere. Theory, Experiments, and Applications. Cambridge, Massachusetts, USA: Academic Press (Elsevier); 2000. pp. 871-942. ISBN 9780122570605

[2] Baumgartner J, Arku RE, Dickinson KL. Household energy solutions in low and middle income countries. In: Encyclopedia of Environmental Health. 2nd ed. Amsterdam, Netherlands: Elsevier; 2019. pp. 494-509. ISBN 9780444639523

[3] World LPG Association. The Role of LGP and biolgp in Europe. Neuilly-sur-Seine, France: WLGPA; 2019

[4] Duy VN, Duc KN, Van NC. Real-time driving cycle measurements of fuel consumption and pollutant emissions of a bi-fuel LPG-gasoline motorcycle. Energy Conversion and Management. 2021;**12**:100135

[5] Usman M, Malik MAI, Ranjha QA, Arif W, Jamil MK, Miran S, et al. Experimental assessment of performance, emission and lube oil deterioration using gasoline and LPG for a sustainable environment. Case Studies in Thermal Engineering. 2023;**49**:103300

[6] Fosudo T, Kar T, Windom B, Olsen D. Low-carbon fuels for spark-ignited engines: A comparative study of compressed natural gas and liquefied petroleum gas on a CFR engine with exhaust gas recirculation. Fuel. 2024;**30**:130456

[7] European Commission. Directorate-general for climate action. In: Going climate-neutral by 2050 – A strategic long-term vision for a prosperous, modern, competitive and climate-neutral EU economy. Brussels: Publications Office; 2019

[8] Raslavičius L, Keršys A, Mockus S, Keršienė N, Starevičius M. Liquefied petroleum gas (LPG) as a medium-term option in the transition to sustainable fuels and transport. Renewable and Sustainable Energy Reviews. 2014;**32**:513-525

[9] Pieter de Jong EAT, Melo SABVD, Mendes-Santana D, Pontes KV. Socio-economic and environmental aspects of BioLPG and bio-dimethyl ether (Bio-DME) production and usage in developing countries: The case of Brazil. Cleaner and Circular Bioeconomy. 2023;**6**:100055

[10] Onwudili JA, Edou DJN. Process modelling and economic evaluation of biopropane production from aqueous butyric acid feedstock. Renewable Energy. 2022;**184**:80-90

[11] Hopwood L, Mitchell E, Sourmelis S. Biopropane: Feedstocks, Feasibility and our Future Pathway. Heslington, York (UK): NNFCC; 2019

[12] Liquid Gas UK. A Practical Approach: Analysis of Off-Grid Heat Decarbonisation Pathways. Warwickshire, UK: Liquid Gas UK; 2019

[13] WLPGA. The Role of LPG and bioLPG in Europe. Neuilly-sur-Seine, France: World LPG Association; 2019

[14] Multiconsult ASA. Annual Report. Oslo, Norway: Multiconsult ASA; 2020

[15] The Global Economy. "theglobaleconomy.com," Business and Economic Data for 200 Countries. 2021. [Online]. Available from: https://www.theglobaleconomy.com/rankings/lpg_consumption/ [Accessed: 18 January, 2024]

[16] United States Environmental Protection Agency. Liquified petroleum gas combustion. In: AP-42: Compilation of Air Emissions Factors from Stationary Sources. Washington DC: USA, EPA; 2008

[17] Expert Market Researh. Europe LPG Market Size, Share, Trends, Forecast. Sheridan, WY, USA: EMR Claight; 2024

[18] European Commission. European Alternative Fuels Observatory: Alternative Fuels. Brussels, Belgium: European Commission; 2024

[19] European Commission. CO_2 Emission Performance Standards for Cars and Vans. Brussels, Belgium: European Commission; 2024

[20] Segarra P. AEGPL Response to Commission's Consultation Strategy to Reduce CO_2 Emissions from Cars. Brussels: European LPG Association; 2016

[21] Ministry of Economic Affairs and Employment of Finland. Reserve Stocks and Preparedness Planning Secure Energy Supply. Helsinki, Finland: Ministry of Economic Affairs and Employment of Finland, Government of Finland; 2024

[22] European Commission Joint Research Centre. Directorate for space, security and migration. In: Liquified Petroleum Gas (LPG) and Liquified Natural Gas (LNG) Sites. Ispra, Italy: EC Publications Office; 2018

[23] Yoshizawa Y. Germany's Biogas Industry and the Prospective Business Opportunities. Use ob Giomethane and bioLNG to Grow in the Transportation Sector. Tokyo, Japan: Mitsui&Co; 2022

[24] Abramiuk-Lété E. LPG Supply Chain in the Context of Energy Supply Security. Brussels, Belgium: Liquid Gas Europe; 2024

[25] Johnson E. Process technologies and projects for BioLPG. Energies. 2019;**12**(2):250

[26] Blue Star Gas. Propane Autogas - The Sustainable Solution [Online]. Available from: https://www.bluestargas.com/autogas/sustainable-fuel/ [Accessed: January 23, 2024]

[27] Autogas. The Autogas market [Online]. Available from: https://www.auto-gas.net/about-autogas/the-autogas-market/ [Accessed: January 25, 2024]

[28] European LPG Association. Autogas in Europe, the Sustainable Alternative. An LPG Industry Roadmap. Brussels, Belgium: AEGPL; 2013

[29] European Alternative Fuels Observatory. Vehicles and fleet. European Commission [En línea]. Available from: https://www.alternative-fuels-observatory.ec.europa.eu/transport-mode/road/european-union-eu27/vehicles-and-fleet [Último acceso: 02 January, 2024]

[30] Liquid Gas Europe. Common Misconceptions about Autogas. Why is this Alternative Fuel Not Being Considered as Part of the Energy Puzzle? Brussels, Belgium: Liquid Gas Europe; 2023

[31] ChemAnalyst. Global LPG Market Analysis Plant Capacity, Production, Operating Efficiency, Demand & Supply, End User Industries, Distribution Channel, Regional Demand, 2015-2030. New York, USA: ChemAnalyst; 2024

[32] Congress of the United States of America. Public Law 102-486-Oct-24, 1992 "Energy Policy Act of 1992". Washington DC, USA; 1992

[33] U.S. Department of Energy. Alternative Fuels Data Center, "Alternative Fueling Station Location" [Online]. Available from: https://www.afdc.energy.gov/stations/#/find/nearest?fuel=LPG&country=US&lpg_secondary=true [Accessed: February 03, 2024]

[34] Munson C, Li C, Durbin TD, Johnson K, Burnette A. On-Road Testing of an LPG Delivery Truck on R-LPG. Riverside, California, USA: University of California; 2020

[35] Propane Autogas. The Growing Presence of Propane in Pupil Transportation. Richmond, Virginia, USA: Propane Education & Research Council; 2014

[36] Autogas. Autogas market trends in the United States. autogas.net [Online]. Available from: https://www.auto-gas.net/government-policies/autogas-incentive-policies/united-states/#:~:text=Autogas%20consumption%20and%20vehicle%20fleet,expanded%20by%205.7%25%20in%202018 [Accessed: February 11, 2024]

[37] USAFacts. What is the most common vehicle fuel type in each state? [Online]. 2023. Available from: https://www.usafacts.org/articles/what-is-the-most-common-vehicle-fuel-type-in-each-state/ [Accessed: February 15, 2024]

[38] Blue Star Gas. State & Federal Autogas Incentives [Online]. Available from: https://www.bluestargas.com/autogas/state-federal-autogas-incentives/ [Accessed: February 17, 2024]

[39] Liu E, Yue S, Lee J. A study on LPG as a fuel for vehicles. Hong Kong: Research and Library Services Division, Legislative Council Secretariat; 1997

[40] Synák F, Čulík K, Rievaj V, Gaňa J. Liquefied petroleum gas as an alternative fuel. Transportation Research Procedia. 2019;**40**:527-534

[41] Twigg MV. Catalyst Handbook. Prescott, Arizona, USA: Wolfe Publishing LTD; 1989

[42] Silva PP, Ferreira RA, Nunes JF, Sousa J, Romanielo L, Noronha F, et al. Production of hydrogen from the steam and oxidative reforming of LPG: Thermodynamic and experimental study. Brazilian Journal of Chemical Engineering. 2015;**32**(3):647-662

[43] Ramantani T, Evangeliou V, Kormentzas G, Kondarides DI. Hydrogen production by steam reforming of propane and LPG over supported metal catalysts. Applied Catalysis B: Environmental. 2022;**306**:121129

[44] Nardo AD, Portarapillo M, Russo D, Benedetto AD. Hydrogen production via steam reforming of different fuels: Thermodynamic comparison. International Journal of Hydrogen Energy. 2024;**55**:1143-1160

[45] Barnoon P, Toghraie D, Mehmandoust B, Fazilati MA, Eftekhari SA. Comprehensive study on hydrogen production via propane steam reforming inside a reactor. Energy Reports. 2021;**7**:929-941

[46] Cu X, Kær SK. Thermodynamic analysis of steam reforming and oxidative steam reforming of propane and butane for hydrogen production. International Journal of Hydrogen Energy. 2018;**43**(29):13009-13021

[47] Parackal B, Khakdaman H, Bourgault Y, Ternan M. An Investigation of direct hydrocarbon (propane) fuel cell performance using mathematical modeling. International Journal of Electrochemistry. 2018;**2018**:5919874

[48] Antolini E. Direct propane fuel cells. Fuel. 2022;**315**:123152

[49] Vishwanathan G. Power Generation: The Emissions Shifting Problem. Washington DC, USA: Propane Education & Research Council; 2022

Chapter 11

Recent Development in LPG Production, Transportation, and Distribution: Focus on Nigeria and African Sub-Regions

Edward Agbai and Edward Aigbedion

Abstract

The global transition towards sustainable and clean energy sources has spurred significant interest in utilizing Liquefied Petroleum Gas (LPG) as a versatile and environmentally friendly energy carrier. This chapter provides a comprehensive overview of recent development in the LPG production, transportation, and distribution in Nigeria and African sub-region. It aligns industry practices driving the adoption of LPG as a pivotal component in energy transition. A significant focus will be placed on technological innovations to enhance LPG production, storage, and distribution. Recent developments in LPG extraction methods, such as advanced refining techniques and the integrating of renewable sources in LPG production including biofuel, are discussed. Policy and regulatory frameworks that support the integration of LPG into national and global energy transition strategies like the Petroleum industry act in Nigeria and the decade of gas policy. The present status of LPG production, its separation methods, environmental implication, challenges, and outlook.

Keywords: clean energy, energy transition, renewable energy, optimization, regulatory framework

1. Introduction

Liquefied Petroleum Gas called LPG is commonly known as propane or butane, has emerged as a versatile and clean-burning energy source with widespread applications in both industrial and domestic sectors. Nigeria, endowed with substantial oil and gas reserves, has been a major player in the global energy market. Historically, the country's primary focus was on crude oil production, leading to underutilization of its vast natural gas resources, including LPG. LPG is gaining importance in Cameroon's energy mix, but supply shortages pose challenges [1] investigated LPG consumption elasticities in Cameroon using annual data from 1994 to 2017. Results show that price, income, and urbanization are significant determinants of LPG consumption.

These findings have implications for demand management and call for policymakers to promote widespread use of LPG, particularly in the savannah zone, to reduce deforestation and biomass dependence. Similarly, Shupler et al. [2] carried out a study in Cameroon, Kenya, and Ghana found that supply-side factors and stove equipment were better predictors of household consumption of liquefied. However, in recent years, there has been a paradigm shift towards recognizing the economic and environmental benefits of LPG. Nigeria Liquified Natural Gas Limited (NLNG) established on May 17, 1989, as a Limited Liability Company. The NLNG was set up to harness Nigeria's vast natural gas resources and produce Liquefied Natural Gas (LNG) and Natural Gas Liquids (NGLs) for export. LPG also known as cooking gas in Nigeria is a very clean versatile and environmentally friendly fuel, an essential commodity for life and living, convenient, portable energy source that is easy to transport and store. It's produced from petroleum refining of crude oil and extraction from natural gas with varying uses ranging from heating purposes, production of aerosol propellant, input to petrochemical industry and as a refrigerant [3, 4]. NLNG has committed to supplying 100% of its LPG production to the domestic market, despite challenges such as feed gas and market challenges. NLNG has supplied over 2.4 million metric tonnes of cooking gas into the domestic market in the last 15 years, spurring a steady rise in yearly consumption. The Federal Government of Nigeria's policies since 2005 have aided supply sustainability in the industry, opening up and broadening the LPG domestic market. The company has dedicated a certain quantity of LPG for the domestic market, generating over 250,000 jobs and creating a foundation for growth by over 1000% [5].

Despite the achievements in the last 15 years, there remains a high demand for cooking gas in Nigeria, outstripping supply, leading to sharp increases in the price of LPG. To mitigate this price surge, the NLNG said the LPG supply chain needs to be improved, with more producers increasing the quantity of LPG on the domestic market and encouraging private entrepreneurs to embark on natural gas processing. The company also called on the Central Bank of Nigeria to ensure the disbursement of loan facilities from a special gas fund (the National Gas Expansion Programme Framework) to competent investors in the LPG value chain. The President of the Nigerian Association of Liquefied Petroleum Gas Marketers (NALPGAM) stated that NLNG remained committed to meeting domestic demand and assured the association that it would continue producing LPG based on the feed gas received from its suppliers. This chapter delves into the dynamic landscape of LPG within the context of the Nigerian petroleum industry and explores its pivotal role in meeting energy demands across the African continent.

2. Overview of LPG production

LPG is obtained through the process of natural gas processing and petroleum refining. The composition of this substance mostly comprises propane and butane, and it is typically stored and transported in a pressurized state in its liquid form [6]. LPG exhibits a notable degree of efficiency and possesses a substantial energy content, rendering it a highly suitable fuel for a wide range of businesses and households on a global scale. The process of producing LPG encompasses a series of distinct stages, which include separation, purification, and fractionation. LPG is a multifaceted energy resource that finds utility in diverse domains such as heating, culinary activities, and transportation [7]. The growing need for cleaner and more sustainable energy sources has underscored the significance of LPG production in addressing these requirements and mitigating carbon emissions. LPG refers to a flammable hydrocarbon fuel that exists in a gaseous state at ambient conditions but may be easily liquefied under moderate pressure [8].

2.1 Economic benefits LPG production

An examination of the economic advantages and employment generation linked to the production of LPG demonstrates that the sector has the capacity to create a significant quantity of jobs directly and indirectly. Below are the following ways LPG plants, its production facilities, and its associated supply chain benefits the Nigerian economy.

2.1.1 Job creation

The establishment and operation of LPG production facilities create job opportunities at various levels, from skilled technicians and engineers to administrative and support staff. Additionally, the distribution and retail aspects of the LPG supply chain also generate employment. The Nigerian Liquefied Petroleum Gas Association (NLPGA) opined that with effective government policy, LPG market can generate over two million jobs [4]. Furthermore, the installation of LPG production facilities necessitates the involvement of proficient workforce in both the construction and operation phases, hence generating employment prospects for the nearby communities. Additionally, the expansion of the LPG sector has the potential to foster the establishment of complementary industries, including transportation and distribution, thereby augmenting employment opportunities [9, 10].

2.1.2 Revenue generation

LPG production contributes to government revenue through taxes, fees, and royalties. The industry can be a significant source of income for the government, helping to fund public services and infrastructure development. With the accelerated LPG penetration, the Nigerian government is targeting a 40% adoption rate (i.e. 13.8 m households) in 5 years by 2026, and 73% adoption in 10 years (33.3 m households) by 2031. The increase in adoption rate will generate revenue to the purse of government [11].

2.1.3 Diversification of the economy

Developing the LPG sector contributes to the diversification of the Nigerian economy. Reducing dependence on traditional sectors like oil and gas can make the economy more resilient to fluctuations in global commodity prices. LPG as the transition fuel provides alternative economic outlet.

2.1.4 Energy access and affordability

Increasing LPG production helps to improve access to clean and affordable energy. As more households and businesses switch to LPG, it reduces reliance on traditional and often more polluting fuels like firewood and charcoal. This transition can lead to cost savings for consumers, as well as health and environmental benefits. With the Cameroonian government encouraging 58% of its population to use LPG as a cooking gas by 2030, up from 20% in 2014. It will provide access to energy for Cameroonian citizens through its National LPG Master Plan designed to scales up the LPG sector to meet this target [12].

2.1.5 Foreign exchange savings

Nigeria currently imports a significant portion of its energy needs. By promoting domestic LPG production, the country can reduce its dependence on imported fuels, leading to savings in foreign exchange.

2.1.6 Indirect economic impact

The use of LPG in various industries and households can result in indirect economic benefits. For example, businesses that rely on affordable and reliable energy for manufacturing processes may experience increased productivity and competitiveness. The productivity will stimulate the growth of local economies and mitigates the prevalence of unemployment and augments income levels, thereby enhancing the overall quality of life for individuals and families.

2.1.7 Environmental benefits

While not directly economic, the environmental benefits of LPG production can have economic implications in the long term. LPG is a cleaner-burning fuel compared to traditional biomass fuels, leading to reduced deforestation, improved air quality, and health-related cost savings. Yeo et al. [13] after investigating the potential economic and environmental benefit of LPG advocated the utilization of LPG by means of governmental policy measures and financial assistance that targets fishing vessels with the strong potential to convert to LPG usages. Such policies are grounded on the efficacy and competitiveness of LPG as a versatile fuel that mitigate air pollutants and limits water pollutants from burning as AGO as a such for fuel for marine activities.

2.1.8 Infrastructure development

The growth of the LPG industry can drive the development of necessary infrastructure, including storage facilities, transportation networks, and distribution channels. This, in turn, can stimulate economic activity in related sectors. According to Witthohn [14] there were about 300 LPG distribution terminals (bottling plants) in the country with a total capacity of about 13,000 tonnes. To fully realize these economic benefits, the Nigerian government and stakeholders must invest in infrastructure, and create an enabling environment for the growth of the LPG sector.

2.2 The environmental implications of LPG production

The world is currently experiencing a significant increase in global energy consumption, coinciding with the urgent need to address the issue of climate change by decreasing carbon dioxide emissions. It is worth noting that a substantial portion of these emissions can be attributed to the utilization of fossil fuels [15]. The assessment of sustainability for LPG production necessitates careful consideration of its environmental impact. The production of LPG is associated with fewer emissions of greenhouse gases in comparison to other forms of fossil fuels, leading to a decrease in the overall carbon footprint [13, 16]. Moreover, the process of LPG burning results in reduced emissions of harmful substances like sulfur dioxide and particulate matter, hence contributing to enhanced air quality and a decrease in respiratory health complications [17]. The environmental advantages render LPG a feasible alternative for facilitating the shift towards a more ecologically friendly and enduring energy framework. This discourse pertains to the examination of the carbon footprint associated with the utilization of LPG in comparison to other fossil fuels. The examination of the carbon footprint of LPG in comparison to other fossil fuels indicates that LPG exhibits a much-reduced carbon footprint. In comparison to coal and oil, LPG demonstrates a reduction of around 20% in carbon dioxide emissions per unit of energy generated [18].

The decrease in carbon dioxide (CO_2) emissions plays a vital role in addressing climate change and achieving international targets for reducing emissions. Moreover, the effective combustion of LPG additionally mitigates the emission of other detrimental greenhouse gases, such as methane, rendering it a compelling substitute for a more environmentally friendly energy trajectory [19]. The exploration and advancement of renewable energy sources, such as solar and wind power, have furthermore played a role in mitigating carbon emissions. Despite this, these sustainable energy sources are less reliable because they do not always work. This makes them less useful for meeting the constant energy needs of both businesses and homes. On the other hand, LPG offers a dependable and conveniently obtainable energy resource that may be utilized for a multitude of purposes, encompassing but not limited to heating, culinary activities, and transportation. LPG is a promising option for facilitating the transition towards a more environmentally sustainable energy landscape, as it boasts a reduced carbon footprint and enhanced combustion efficiency. This alternative energy source holds potential for addressing our energy requirements while simultaneously mitigating the adverse impacts of traditional energy sources on the environment [20].

3. LPG price watch

Based on current statistics, there has been a consistent upward trend in the price of LPG, which can be attributed to the escalating need for environmentally friendly and sustainable energy alternatives. The demand for LPG in Nigeria has experienced steady growth, driven by increasing urbanization, population expansion, and a growing awareness of the benefits of clean cooking fuels. The domestic sector, particularly households and small businesses, has emerged as the primary consumer of LPG. Efforts to encourage LPG adoption through subsidies, awareness campaigns, and targeted government programs have played a crucial role in shifting consumer preferences towards this cleaner alternative to traditional biomass fuels.

Refilling a 5 kg LPG cylinder according to [21], cost ₦3594.81 an equivalent of $4.49 using the conversion rate of 800 Naira to 1US Dollars in December 2021, an 84.37% increase from December 2020 which was ₦1949.75. Average monthly price rose 8.53% from ₦3312.42 in November 2021. The states of Benue, Cross River, and Borno had the highest average prices for refilling a 5 kg LPG cylinder, while the states of Adamawa, Yobe, and Bauchi had the lowest, with prices of ₦2398.40, ₦2543.14, and ₦2586.43, respectively. As of December 2021, the average cost of refilling a 12.5 kg LPG cylinder rose by 76.49% to ₦7332.04 from ₦4154.28 in December 2020. The average price rose 0.33% from ₦7308.06 in November 2021 per month-on-month study. The highest average prices for refilling a 12.5 kg LPG cylinder were in the states of Osun, Oyo, and Abuja the Federal Capital Territory with the sum of ₦ 8491.67, ₦8303.33, and ₦8058.00 respectively, while the states of Borno, Bayelsa, and Nasarawa had the lowest average prices of ₦5852.13, ₦6678.57, and ₦6679.57 respectively [22].

Many nations in Africa have acknowledged the advantages of LPG and have enacted legislation and provided incentives to promote its production and utilization. Nevertheless, there are persistent issues that need to be addressed to guarantee a steady and dependable provision of LPG, particularly in less developed areas. There is a concerted endeavor to enhance the infrastructure for LPG and optimize the distribution networks to effectively address the increasing worldwide demand.

4. Transportation of LPG in Nigeria and West Africa

LPG is very portable. There are several ways to transfer it, most commonly in liquid form, because its volume is lower than that of gas. Large tankers transport LPG overseas and large gas transporters typically provide this function. There are intermodal tank containers that can transport LPG, these stackable containers enable the transportation of LPG by ship, rail, or truck, eliminating the need to empty and reload. Tanker rail cars can transport vast quantities of LPG. Tanker rail cars frequently deliver LPG from the well sites to the terminals. Transporting goods between the two terminals requires expertise. Transferring LPG from gas sources to storage terminals occasionally requires the use of designated pipelines. The prohibitive expense of building them prevents their extensive use. In some areas, LPG reticulation is prevalent. LPG is distributed as reticulated gas through a network of pipes. The pipes are linked to a storage container located at a distance from the customer's location, and several customers are supplied by a single storage container. Mode Investment and Development Services (MIDs) Gas Group manages the distribution systems for LPG in Nigeria and West Africa. Gas Africa, a company based in Nairobi, Kenya, provides the service of installing LPG reticulated piping systems in standalone buildings. This includes setting up centralized cylinder banks in residential apartment blocks and installing bulk LPG tanks in hotels. Implementing a pre-paid metering system for residential use [23, 24].

5. LPG distribution in Nigeria and African sub-region

LPG is a cleaner cooking fuel widely promoted in many countries. Interventions to promote LPG have varied outcomes, with some studies showing positive effects and others showing ineffectiveness. The Rural LPG Promotion Program (RLPGPP) in Ghana, launched in 2013, aimed to make LPG the primary cooking fuel for rural residents by providing free LPG cylinders, cookstoves, and accessories. However, the program's effect remains to be measured despite being discontinued. Adjei-Mantey and Takeuchi [25] carried out a study to examines the factors influencing the adoption of clean cooking fuels in Ghana, focusing on the supply-side determinants of LPG usage. The research found that distance to refill stations negatively affects LPG adoption and usage frequency, but multiple services at refill stations increase usage. Other key factors include policy interventions, behavioral and socio-economic characteristics of households. The findings have significant policy implications for developing countries facing infrastructure challenges.

LPG is a crucial commodity in Nigeria, primarily produced by the NLNG. It can be transported in various ways, including ship, rail, tanker trucks, intermodal tanks, cylinder trucks, pipelines, and local gas reticulation systems. Distribution logistics in Nigeria are a major challenge, with inadequate facilities such as small low draft vessels, channel draft restrictions, and transportation infrastructure contributing to increased landing costs. The leading producer of LPG in Nigeria is the NLNG in its bonny plant. It is transported to Lagos, where operational storage facilities are located. The current distribution network in Nigeria has a single functional outlet in Lagos, with limited capacity in coastal cities like Port Harcourt and Calabar [26].

The optimization routine study by [27] was found to reduce trucking costs of LPG using specific distribution outlets, compared to the current trend of using only the Lagos distribution outlet. The model showed that distributing LPG from the proposed Port Harcourt channel to Kano and Sokoto is more cost-effective (24% reduction) than transporting it from Lagos. Distributing LPG from Calabar to Gombe and Maiduguri is more profitable (28% reduction) than from Lagos. The most cost-effective distribution outlets for LPG transportation to specific cities across the country are the proposed Calabar and Port Harcourt outlets. The LPG distribution network from the Lagos outlet remains strategic for distribution to cities closer to it.

Lasisi [28] carried out a study to examines the production and distribution of LPG in Nigeria, focusing on historical and predictive data from 1994 to 2020. the result of the study indicated that LPG consumption increased rapidly from 2016 upward, driven by infrastructural surplus, investment, and awareness initiatives. The results suggest a strong correlation between demand-supply bonds driven by market forces in urban and suburban cities, providing reliable parameters for sustainable strategic planning and policy development for economic recovery.

6. Challenges, outlook, and Nigerian decade of gas policy

6.1 Challenges

6.1.1 Infrastructure deficiency and limited domestic production

Inadequate LPG storage, transportation, and distribution infrastructure poses a significant challenge. A well-established distribution network is necessary to supply LPG to consumers efficiently. Also, Nigeria relies on imported LPG to augment the supply from NLNG, contributing to supply chain vulnerabilities and price fluctuations. Increasing domestic production could enhance supply security [29].

6.1.2 Safety concerns

Safety issues related to the handling and use of LPG and its cylinders can deter consumers. On the production side of the value chain, the volatility of the gas portends safety issues [30]. Proper safety measures, public awareness campaigns, and regulatory enforcement are necessary to address these concerns.

6.1.3 Pricing and subsidy issues and lack of regulatory framework

Fluctuations in global oil prices can impact the cost of LPG in Nigeria. Subsidy challenges and price variations may affect the affordability and accessibility of LPG for consumers. Inconsistent regulatory frameworks and enforcement can lead to sub-optimal industry performance. A clear and robust regulatory environment is essential for the sustainable growth of the LPG sector [31].

6.2 Outlook

6.2.1 Infrastructure development and diversification of supply sources

Investing in infrastructure expansion, including storage facilities, pipelines, and distribution networks, is crucial. Developing a comprehensive and efficient infrastructure will facilitate the seamless distribution of LPG. Reducing dependency on imported LPG by encouraging domestic production and exploring alternative supply sources can enhance supply stability and reduce the impact of global price fluctuations [32].

6.2.2 Consumer education and awareness

Public awareness campaigns on the safety and benefits of using LPG can help overcome misconceptions and increase consumer acceptance. This includes educating consumers on the proper handling, storage, and usage of LPG.

6.2.3 Government incentives and international collaboration

Incentivizing and supporting investors and stakeholders in the LPG sector can stimulate growth. This may include tax incentives, subsidies, and other measures to encourage production, distribution, and retail investment. Also, international organizations collaborating with other countries with successful LPG programs can provide valuable insights and support for Nigeria's LPG industry development [33].

6.3 Nigeria's decade of gas policy 2021

Nigeria has been actively pursuing policies related to the oil and gas sector, which includes natural gas. Historically, Nigeria has been a significant player in Africa's oil and gas industry. With its vast natural gas reserves, the government has developed policies to optimally utilize these resources for economic growth and diversification.

In March 2021, former Nigerian President Muhammadu Buhari declared the 2020s as Nigeria's 'decade of gas.' In doing so, Nigeria signaled a renewed focus on gas as the fuel of choice for powering industrial ambitions. In 2022, natural gas demand in Nigeria reached around 12.12 billion standard cubic meters, up from approximately 10.11 billion standard cubic meters in 2021 [34]. In establishing its position as the tenth largest gas reserve country, the decade of gas policy is intended to reduce gas flaring. Gas flaring is a common practice where associated gas is burned off during oil extraction. The policies aim to reduce environmental impact and encourage the capture and utilization of associated gas. The policy also encourages gas infrastructure development in the transportation and distribution through pipelines, gas processing facilities, and liquefied natural gas (LNG) projects.

The decade of gas policy also aims to increase domestic gas utilization and open the sector for investment and collaboration by promoting the use of natural gas for power generation, industrial processes, and household consumption [35]. Attracting investment in the gas sector through collaboration with international partners for technology transfer and expertise as an added strategy.

Another impact area of the decade of gas policy is robust framework and regulatory compliance. Establishing an improved regulatory framework was crucial for effectively implementing gas policies for licensing, pricing, and environmental standards. Implementing and enforcing clear policies and regulations can opportunities for private sector participation and investment while complying with safety standards and environmental regulations is essential.

For the decade of gas as it relates to LPG, integrating LPG into Nigeria's national and global energy transition strategies requires robust policy and regulatory frameworks like:

6.3.1 Financial incentives

Subsidies, tax incentives and investment guarantees were instrumental to the policy adoption of the LPG subsector. Subsidies will help promote the adoption of LPG and make it more competitive compared to other energy sources. The policy also offers guarantees or insurance mechanisms to attract investments in the LPG sector and mitigate risks for investors.

6.3.2 Consumer awareness and education

Public awareness campaigns and training programs will be vigorously pursued. The policy made provisor for public awareness campaigns to educate consumers about the benefits of LPG, safety measures, and its role in the energy transition. It will also implement training programs for practitioners, businesses, and consumers to enhance understanding and encourage the safe use of LPG.

6.3.3 Monitoring and evaluation

Regular assessments and data collection are encouraged. Implement a system for regular assessments of the effectiveness of LPG policies and regulations, with adjustments made to achieve desired outcomes. Also, mechanisms were established for collecting and analyzing data on LPG usage, safety incidents, and market trends to inform policy decisions.

7. Conclusion

LPG is a promising alternative for addressing global energy security and reducing carbon dioxide emissions, particularly in rural regions where alternative energy options may be scarce. The production of LPG reduces greenhouse gas emissions by around 20% per unit of energy generated, making it a more environmentally friendly alternative to coal and oil. LPG also contributes to improved air quality and reduced respiratory health complications.

In Nigeria and African sub-region, LPG is transported through various means, including ship, rail, tanker trucks, intermodal tanks, cylinder trucks, pipelines, and local gas reticulation systems. However, distribution logistics in Nigeria are challenging due to inadequate facilities, channel draft restrictions, and transportation infrastructure. There is a concerted effort to enhance the infrastructure for LPG to optimize distribution and transportation of LPG, the most cost-effective distribution outlets for LPG transportation are the proposed across African sub-region. Despite significant progress, challenges persist, including infrastructure constraints, pricing fluctuations, and the need for continued regulatory refinement. Environmental considerations and the push for sustainable energy solutions will shape the trajectory of the LPG sector in Nigeria. And African sub-region.

Author details

Edward Agbai[1*] and Edward Aigbedion[2]

1 School of Entrepreneurship, Emmaneul University, Raleigh, NC, USA

2 Emerald Energy Institute, University of Port Harcourt, Port Harcourt, Rivers State, Nigeria

*Address all correspondence to: edwardagbai@gmail.com

IntechOpen

© 2024 The Author(s). Licensee IntechOpen. This chapter is distributed under the terms of the Creative Commons Attribution License (http://creativecommons.org/licenses/by/4.0), which permits unrestricted use, distribution, and reproduction in any medium, provided the original work is properly cited.

References

[1] Sapnken FE, Kibong MT, Tamba JG. Analysis of household LPG demand elasticity in Cameroon and policy implications. Heliyon. 2023;**9**(6):1-14. DOI: 10.1016/j.heliyon.2023.e16471

[2] Shupler M, Mangeni J, Tawiah T, Sang E, Baame M, Anderson de Cuevas R, et al. Modelling of supply and demand-side determinants of liquefied petroleum gas consumption in peri-urban Cameroon, Ghana, and Kenya. Nature Energy. 2021;**6**(12):1198-1210

[3] Omeje I. Nigeria: Govt LPG Penetration Agenda Faces Challenges As Price Soars. Major Waves Energy Report. 2023. Available from: https://www.majorwavesenergyreport.com/nigeria-govt-lpg-penetration-agenda-faces-challenges-as-price-soars/

[4] Paczuski M, Marchwiany M, Puławski R, Pankowski A, Kurpiel K, Przedlacki M. Liquefied petroleum gas (LPG) as a fuel for internal combustion engines. Alternative Fuels, Technical and Environmental Conditions. 2016;**13**:105-136. DOI: 10.5772/61736

[5] Adekoya F. NLNG's Intervention in Bridging LPG Supply Gap. 2023. Available from: https://guardian.ng/energy/nlngs-intervention-in-bridging-lpg-supply-gap/#:~:text=NLNG%20said%20it%20currently%20supplies,journey%20of%20domestic%20gas%20supply

[6] Johnson E. Process technologies and projects for bio-LPG. Energies. 2019;**12**(2):250. DOI: 10.3390/en12020250

[7] Falkiner RJ. Liquefied petroleum gas. In: Fuels and Lubricants Handbook. West Conshohocken, PA, USA: ASTM International; 2003. pp. 31-59

[8] Yezid Rozo-Álvarez C, Álvarez-Pomar L, Gómez-Vargas E, Lizcano-Caro I. The liquefied petroleum gas, a sectorial analysis. Visión Electrónica. 2019;**13**(2)

[9] Kojima M. The role of liquefied petroleum gas in reducing energy poverty. 2011. Available from: https://documents1.worldbank.org/curated/en/390661468157779954/pdf/674520NWP00PUB0084180LPG0Report0Web.pdf

[10] Tamba JG. LPG consumption and economic growth, 1975-2016: Evidence from Cameroon. International Journal of Energy Sector Management. 2021;**15**(1):195-208. DOI: 10.1108/IJESM-01-2020-000

[11] Nigeria LNG. Who We are. 2023. Available from: https://www.nigerialng.com/the-company/Pages/Who-We-Are.aspx

[12] Kypridemos C, Puzzolo E, Aamaas B, Hyseni L, Shupler M, Aunan K, et al. Health and climate impacts of scaling adoption of liquefied petroleum gas (LPG) for clean household cooking in Cameroon: A modeling study. Environmental Health Perspectives. 2020;**128**(4):047001

[13] Yeo SJ, Kim J, Lee WJ. Potential economic and environmental advantages of liquid petroleum gas as a marine fuel through analysis of registered ships in South Korea. Journal of Cleaner Production. 2022;**330**:129955

[14] Witthohn R. Raw materials. In: International Shipping: The Role of Sea Transport in the Global Economy. Wiesbaden: Springer Fachmedien Wiesbaden; 2022. pp. 135-250

[15] Ikealumba WC, Wu H. Some recent advances in liquefied natural gas (LNG) production, spill, dispersion, and safety. Energy & Fuels. 2014;**28**(6):3556-3586. DOI: 10.1021/ef500626u

[16] Kolb S, Plankenbühler T, Hofmann K, Bergerson J, Karl J. Life cycle greenhouse gas emissions of renewable gas technologies: A comparative review. Renewable and Sustainable Energy Reviews. 2021;**146**:111147

[17] Ryskamp R. Emissions and Performance of Liquefied Petroleum Gas as a Transportation Fuel: A Review. Neuilly-sur-Seine, France: World LPG Association; 2017. Available from: https://auto-gas.net/wp-content/uploads/2019/11/2017-WLPGA-Literature-Review.pdf

[18] Van Leeuwen R, Evans A, Hyseni B. Increasing the Use of Liquefied Petroleum Gas in Cooking in Developing Countries. Live Wire; 2017/74. Washington: World Bank; 2017

[19] Murshed M, Ali SR, Banerjee S. Consumption of liquefied petroleum gas and the EKC hypothesis in South Asia: Evidence from cross-sectionally dependent heterogeneous panel data with structural breaks. Energy, Ecology and Environment. 2021;**6**:353-377. DOI: 10.1007/s40974-020-00185-z

[20] Rivera XCS, Topriska E, Kolokotroni M, Azapagic A. Environmental sustainability of renewable hydrogen in comparison with conventional cooking fuels. Journal of Cleaner Production. 2018;**196**:863-879. DOI: 10.1016/j.jclepro.2018.06.033

[21] Neeka JB, Ikoku U, Iyalla E, Joel O, Ikiensikimama S. Predictive models on viable options for liquefied petroleum gas LPG distribution: Case for a small and medium scale enterprise SME commodity in Nigeria. In: SPE Nigeria Annual International Conference and Exhibition (Eko Hotels and Suites, Lagos, Nigeria). OnePetro; Aug 2019

[22] National Bureau of Statistics. Liquefied Petroleum Gas (Cooking Gas) Price Watch. 2021. Available from: https://nigerianstat.gov.ng/elibrary/read/1241117

[23] Clarkson Plc. How is LPG Transported. 2023. Available from: https://www.clarksons.com/glossary/lpg-transport/

[24] Gas Africa. LPG Reticulated Piping System. 2024. Available from: https://gasafrica.com/services-lpg-reticulated-piping-system

[25] Adjei-Mantey K, Takeuchi K. Supply-side factors of LPG adoption and usage frequency in Ghana: Assessing the validity of subjective distance to refill. Energy for Sustainable Development. 2022;**70**:475-481. DOI: 10.1016/j.esd.2022.08.021

[26] Croitoru L, Miranda JJ, Khattabi A, Lee JJ. The Cost of Coastal Zone Degradation in Nigeria: Cross River, Delta and Lagos States. Washington, DC: World Bank Group; 2020

[27] Ihemtuge TU, Aimikhe VJ. Optimization of liquefied petroleum gas (LPG) distribution in Nigeria. International Journal of Engineering and Technology Research. 2020;**10**:8-14

[28] Lasisi S. An analysis of liquefied petroleum gas adoption for household cooking in Nigeria [doctoral dissertation]. Walden University; 2021

[29] Matthews WG. Opportunities and challenges for petroleum and LPG markets in sub-Saharan Africa. Energy Policy. 2014;**64**:78-86

[30] Cornell PE. Energy and the three levels of national security:

Differentiating energy concerns within a national security context. Connect. 2009;**8**(4):63-80

[31] Dhali M, Hassan S, Subramaniam U. Comparative analysis of oil and gas legal frameworks in Bangladesh and Nigeria: A pathway towards achieving sustainable energy through policy. Sustainability. 2023;**15**(21):15228

[32] Mac Kinnon MA, Brouwer J, Samuelsen S. The role of natural gas and its infrastructure in mitigating greenhouse gas emissions, improving regional air quality, and renewable resource integration. Progress in Energy and Combustion Science. 2018;**64**:62-92

[33] Singh D, Pachauri S, Zerriffi HJERL. Environmental payoffs of LPG cooking in India. Environmental Research Letters. 2017;**12**(11):115003. DOI: 10.1088/1748-9326/aa909d

[34] Yin Y, Lam JSL. Bottlenecks of LNG supply chain in energy transition: A case study of China using system dynamics simulation. Energy. 2022;**250**:123803

[35] Adewuyi OB, Kiptoo MK, Afolayan AF, Amara T, Alawode OI, Senjyu T. Challenges and prospects of Nigeria's sustainable energy transition with lessons from other countries' experiences. Energy Reports. 2020;**6**:993-1009

Chapter 12

The Impact of BioLPG's on Carbon Footprint: A Context of European Countries

Amir Sepehri and Mohammad Kamal Ghassem Alaskari

Abstract

There is an urgent requirement for transition to better sustainable bioeconomy techniques due to global worries about the depletion of the fossil fuel supply in the world. Alternative fuels have gained interest as the world strives to create sustainable energy sources and reduce greenhouse gas emissions. BioLPG, a renewable shape of Condensed Petroleum Gas [LPG], has emerged as an attractive option in response to energy demands and environmental concerns. Using biomass feedstocks, such as agricultural residues, forestry waste, and waste cooking oil, bioLPG is a carbon-neutral alternative to traditional LPG. Compared with conventional LPG, bioLPG offers several benefits, including reduced net greenhouse gas emissions and lower carbon intensity. Moreover, bioLPG production can utilize multiple biomass feedstocks and maximize waste value. This chapter assesses the current state of research on bioLPG, identifies critical challenges and issues, and presents potential solutions for the broad adoption of bioLPG. BioLPG's footprint varies and depends on the feedstock and situation in the European bases. However, it is often eligible for government support through financial credits and meets biofuel requirements by EU countries under the Renewable Energy Directive.

Keywords: carbon footprint, greenhouse gas emissions, renewable, bioLPG, carbon intensity, biomass feedstocks

1. Introduction

There is an urgent requirement to transition to better sustainable bioeconomy techniques due to global worries about depleting fossil fuel supply [1, 2]. The worldwide hunt for alternative fuels is expanding due to concerns about climate change and a desire to lessen reliance on fossil fuels. It is also possible for liquefied petroleum gas [LPG], a fuel combination mostly made up of Propane and butane in different ratios [3], to become a partly renewable fuel if it is made from feedstocks such as vegetable oils, biomass, and oily wastes. Proponents of decarbonization and the LPG sector are keen to increase the amount of BioLPG produced. The hydro treatment of vegetable oils (HVO) pathway, which involves hydrogenating the triglycerides contained in that raw vegetable material, is a technologically proven method for this to happen sustainably [4]. The method yields Propane [bioLPG] as a byproduct and HVO biodiesel (commonly known as "green" diesel as the primary product when carried out under carefully regulated temperature and pressure conditions [5]. From this angle, bioLPG appears to be a promising drop-in fuel—that is when replacing fossil LPG, it does not require any modifications to the infrastructure or equipment that will be used [6], and it also has fewer adverse environmental effects related to energy generation [7].

Research on the potential function of bioLPG in decarbonizing transportation still needs to be done. It has been proved that LPG may be used in a proton-exchange membrane fuel cell to provide supplementary power for transportation. Net-zero carbon transition options that economically sustainably decarbonize current energy routes provide significant advantages when meeting the growing and pressing demands of mitigating climate change worldwide. As the world moves toward a circular economy, producing bioLPG from renewable resources would be a profitable way to deliver climate-friendly fuel to LPG distributing and user ecosystems, which are currently present in almost every nation and can grow [8]. In the current global effort to develop renewable energy sources and lower greenhouse gas emissions, alternative fuels are becoming more and more attractive. BioLPG, a renewable version of liquefied petroleum gas, has become a desirable alternative in response to environmental concerns and energy needs. It is crucial to ascertain the effects of the fuel of interest on the environment, particularly about carbon footprint. This chapter aims to determine how bio LPG affects carbon footprint. It evaluates the status of bioLPG research as it is now, highlights essential problems and obstacles, and offers possible solutions to promote the widespread use of bioLPG.

2. Production of BioLPG

Using commercial procedures employing HydroDeoxygenation (HD) reactors, including Universal Oil Products (UOP's) Eco-fining technique and Neste Renewable Diesel Process (NRDP), a range of (lower-molecular weight) (LMW) biofuels, comprising bioLPG, are being created from vegetable oil and animal fats [9]. With a biodiesel:bioLPG output ratio of around 9–10:1 in weight, bioLPG is often recovered as byproducts from the manufacture of hydrotreated vegetable oil (HVO) or Hydrotreated Esters and Fatty Acids (HEFA) biodiesel [10]. Triglycerides are found in common fats and oils [11]. A three-carbon backbone is joined via ester linkages to three long-chain hydrocarbons. By adding hydrogen to the ester to change its oxygen to water, the HVO process disrupts these bonds. It splits one triglyceride molecule into three long-chain hydrocarbon molecules, each with an average of 16–18 carbons, and one propane molecule [C3H8]. The hydrogenation of natural fats and oils into biodiesel and bioLPG in the HVO process is shown in the equation below [12].

$$\mathrm{C_{57}H_{104}O_6 + 12\,H_2 \rightarrow C_{17}H_{36} + 2\,C_{18}H_{38}} + CO_2 + \mathrm{4\,H_2O + C_3H_8} \quad (1)$$

The Eco-fining process for the co-production of green diesel and bioLPG from the hydrogenation of vegetable oil is shown in **Figure 1**. The vegetable oil is hydrogenated in the reactor, and the resulting products are separated into water and carbon dioxide through HydroDeoxygenation (HD) and DeCarboxylation (DC), respectively. The hydrocarbons are then fractionated by distillation to obtain the bioLPG, Naphtha, and diesel [13].

DOI: http://dx.doi.org/10.5772/intechopen.1004239

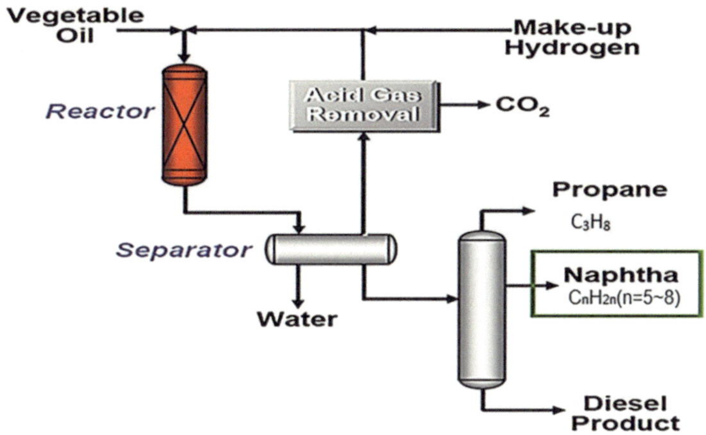

Figure 1.
The complete process of the bioLPG component.

Through the HVO procedure, hydrogen reacts with a triglyceride vegetal (or animal) oil to produce sustainable diesel, an aliphatic hydrocarbon having 16 to 18 carbons. Other reaction products include carbon dioxide, water, and a combination of smaller hydrocarbons, namely Propane. Within the HVO procedure, processes such as decarboxylation (DCO) and hydrodeoxygenation (HDO) take place. One reaction may be preferred by modifying the catalyst and process parameters, albeit a combination of the two will always occur. Preferring DCO or HDO significantly alters the outputs' makeup. According to [14], the theoretical results for the input of brown grease include 15.5% weight% CO_2 for DCO and 12.7% weight% H_2O hydrodeoxygenation (HD). The hydrogen ions will remain unsaturated regardless of the chosen option since hydrogen is needed for the saturation of the olefin carbon linkages.

3. Impact of bioLPG on carbon footprint

After analyzing the carbon emissions of BioLPG at the points of production and consumption, the influence on carbon footprint may be determined. Its emissions during consumption are similar to that of fossil LPG; it is chemically comparable to fuel and could be safe for human health [8]. Islam et al. [15] report that the highest carbon emissions in combustion are discovered at $\Phi = 2.32$ for the steady-state scenario of BioLPG consumption within a high-pressure liquid rocket powering system. At $\Phi = 1.50$, the maximal specific impulse is 327.92. Fortunately, lean mixture utilization is preferred for clean fuel combustion and a lower carbon impact. With $\Phi = 1.50$, the carbon release is 41%, a much higher value for this process. Conversely, the CO emission from the lean blend at $\Phi = 0.80$ at the nozzle exit is relatively low, at 14.7%. Having 310.96 for $\Phi = 0.80$, developing power is 57.73 MW, comparable to that found for $\Phi = 1.50$. The optimal combustibility of BioLPG in $\Phi = 0.80$ is suitable for environmentally friendly clean burring.

Processing bioLPG has an adverse environmental impact but to a lesser extent than that of its non-renewable competitors. Therefore, methodically examining the fuel's environmental performance, identifying any possible effects, and suggesting remedies for its reduction or eradication is an effective strategy for handling this issue [16].

Because of its systematic and quantitative qualities, The life-cycle assessment (LCA) approach can definitively conduct such an evaluation [17].

LCA research on biofuels generated by the HVO approach has been documented in the literature for the last 15 years; however, they are rare [18]. A screening life cycle assessment (LCA) was used to assess the environmental effects of diesel produced by hydrotreating Jatropha oil versus diesel produced using traditional technology that consumes crude oil. Various processing conditions for agricultural produce, soil types, and byproduct utilization are considered when assessing the alternate route. In terms of usage of resources, global warming, and summertime smog, the HVO process for Jatropha fared better than petroleum diesel. However, it did worse in terms of acidification and eutrophication. According to [18], greenhouse gas (GHG) dynamics depend on biomass and carbon stocks as vegetation cover changes. GHG emissions are also influenced by modifications in soil transformation brought about by the establishment of jatropha crops. Researchers have also emphasized how farming on low-carbon soils has a tremendous potential to mitigate global warming. [19] conducted a comparable, however more comprehensive, study in which they contrasted the production of renewable diesel derived from palm oil, rapeseed, and Jatropha to that of fossil fuel. Their findings support the findings of [18] since, for every feedstock they looked at, the "green" energy had a much lower global warming potential (GWP). [20] carried out an LCA assessment of HVO biodiesel produced in an oil refinery's hydrotreatment unit by co-processing soy oil with traditional fossil fuel. In such a scenario, the predicted environmental advantages of the mix encompass a primary energy demand (PED) decrease of as much as 2.0% and a GWP decline of 9.0% when compared to the energy purely fossil, even with the assumption of 13% v/v renewable blend.

According to the publications mentioned above, life cycle studies on the HVO approach concentrate on producing "green diesel," with bioLPG as a byproduct of this process. On the other hand, used the opposite approach and calculated the environmental impact of biopropane HVO made from plant-based inputs (palm oil, rapeseed, palm fatty acid distillate (PFAD), and soy oil) as well as one animal source (tallow, used cooking oil: UCO). The life cycle assessment (LCA) was conducted using a "cradle-to-gate" perspective, considering 1.0 MJ of energy related to the bioLPG Functional Unit. However, they are restricted to the impact of global warming. The consequences ranged from 5.2 g CO2/MJ—for Used Cooking oil and taking into account energy allotment at the HVO unit—to 102 g CO2/MJ—for rapeseed oil, including indirect land utilization changes as well as energy allocation [12]. The author concluded that this spectrum of effects resulted from the kind of input the HVO technology uses and how this input influences the operating environment.

They found [12] that when the model is run using palm oil substrate to make Renewable energy, the output is 40 g CO2/MJ. Its main components are methane from the decomposition of palm oil mill effluents (POME), carbon emissions from gas-intensive activities that produce hydrogen and nitrogenous fertilizer, and carbon emissions from palm oil plantation activities (mostly from logistic vehicle emissions). The span of "official" estimations for this finding is 39 to 50 g CO_2/MJ. Outside of the stated range, there are other approximations. According to research by Eco-invent, for example, the footprint of palm oil alone is estimated to be 42 g CO_2/MJ. This likely results in a renewable diesel footprint of more than 50 g CO_2/MJ when HVO emissions are accounted for by [12]. In Ref. [21] estimated that palm oil footprint is 241 g CO_2/MJ. This seems significantly more comprehensive than a footprint computed using traditional techniques and likely contains a significant indirect land use change (Ilic) component [22].

The HVO biopropane footprint is determined in [12], applying economic allocation and the calibrated footprint method, resulting in to16 g CO_2/MJ. This represents the first accurate, publicly available footprint assessment for HVO biopropane [23]. Research conducted in 2014 estimates CO_2/MJ to be between 10 and 50 g; however, it did not provide a base scenario [24]. The HVO bio LPG footprint varies significantly under different estimated conditions. The bioLPG footprint increases by about 16 g, and the biodiesel footprint falls somewhat when the allocation is changed from economics to energy. Additionally, this alters the oil mill's footprint. By including methane collection, the footprint is sunk by over 7 g. Then, there is the issue of indirect land use change (ILUC). An (ILUC) magnitude of 55 CO_2/MJ is suggested for "oil crops" in Annexes V and VIII of the December 17, 2012 Modification to the EU Renewable Energies Act and the Fuel Quality Act. Many questions still need to be answered about whether (ILUC) should be incorporated and its criteria. They reported that the range of presently suggested parameters for palm oil is 44–231 CO_2/MJ.

Does crude HVO bio LPG footprint fall within the residual category? This makes sense since glycerine from manufacturing Fatty Acid Methyl Esters (FAME) biodiesel is categorized as a residue within Research, Evaluation, and Development RED. It is produced in a manner comparable to HVO propane, meaning it is an inevitable result of the synthesis of biodiesel [25]. Nevertheless, the UK Government has decided that HVO bioLPG is a co-product and not a residue, expressly rejecting this designation [26]. The HVO bio LPG footprint could be around 8 CO_2/MJ if crude HVO propane is considered a residue [12]. Additionally, there is the possibility that palm oil may be categorized as a residue or trash.

The footprint for the palm fatty acid distillate (PFAD) feedstock economic-allocation example is 15 g CO_2/MJ, with the majority of its elements being the same as those for HVO biodiesel, naturally. Depending on the circumstance, its footprint varies between 5 g and 80 g CO_2/MJ. The economic allocation for tallow yields a footprint of 17 g CO_2/MJ. Its main components are the HVO process and the relatively energy-intensive tallow manufacturing [27]. Tallow significantly reduces the footprint if it is considered a residue. It decreases to 5 g CO_2/MJ with economic provisioning, while the footprint increases to 11 g under energy allotment. Rapeseed oil, a popular frying oil in Europe, is presumed by utilized cooking oil to have been the oil that was used in the fryer. According to reference [12], the utilized cooking oil footprint in the economic-allotment scenario is 9 g CO_2/MJ. Its main components are producing hydrogen, rapeseed, nitrogenous fertilizer, and steam (from the oil mills and HVO reactor). The footprint decreases to 5 g over economic allotment and 11 g over energy allotment if utilized cooking oil is categorized as a residue. Under the economic-allotment scenario, the footprint of rapeseed oil is 19 g CO_2/MJ. Its main components are the HVO process, the manufacturing of nitrogen fertilizer, and rapeseed. If energy is allocated instead of economic significance, the footprint increases to 47 g CO_2/MJ. Should an iLUC component be used in addition, the footprint increases to 102 g. In the economic scenario, the footprint of soybean oil is 17 g CO_2/MJ. Its main components include soybean cropping, hydrogen generation, and the HVO mechanism. If energy is allocated instead of economic benefit, the footprint increases to 40 g CO_2/MJ. If an iLUC component is put on top of it, the footprint increases to 95 g CO_2/MJ.

This technique considered the environmental impact of bioLPG on carbon footprint varies substantially based on the feedstock and circumstances. However, it is frequently qualified for government support through financial credits and mandates for biofuels issued by EU members in compliance with the Renewable Energy

Directive [28]. According to the footprint calculation, there are documented instances of this footprint fluctuation, including research on forklifts [29] and products made from forest products [30]. Among the footprint users who should be aware of this potential volatility are regulators, suppliers, and consumers.

4. BioLPG projects and benefits in Europe context

Currently manufacturing bioLPG in Europe are the following companies: Repsol (Spain), Neste (the Netherlands), Global Bioenergies (France), Eni (Italy), and Global Bioenergies (France). According to estimates, the annual consumption of branded bio LPG—a product clearly labeled as such and accessible on the market—was around 100 kilotonnes in 2018. Today, the remaining 100 kilotonnes of bioLPG generated annually are used internally as process fuel5. BioLPG is present in modest but constantly increasing amounts. BioLPG is accessible in various European markets, including the Netherlands, Germany, Sweden, Denmark, France, Ireland, the United Kingdom, and Belgium.

Renewable liquid gas provides a long-term, affordable option to cut carbon emissions and air pollutants from difficult-to-decarbonize industries like transportation and rural heating. It is backed by regulations that encourage R&D efforts and innovative production methods. Among the cleanest fuels on the market is LPG, especially compared to traditional, high-carbon fuels like coal, heating oil, diesel, and gasoline. CO_2 outflows can be reduced by up to 55% while utilizing LPG and by as much as 83% after adopting bioLPG and replacing an oil boiler with an LPG boiler [31]. In addition, compared to other energy sources, LPG-derived renewable resources offer a substantial potential to lower air pollution. Ordinary warming frameworks, such as gas boilers and combined heat and power (CHP) units, may also smoothly employ LPG. Due to its compatibility with current heating systems, businesses may see a reduction in investment costs and a smoother transition from fossil fuels with more significant carbon emissions to LPG and bioLPG, equal to renewable energy. Modern, cutting-edge heating systems, including hybrid and gas-driven heat pumps, may be powered by LPG in commercial and industrial settings. LPG boilers produce 80–99% less PM and 50–75% less nitrogen oxide (NOx) than solid or liquid fuels (such as biomass, peat, heating oil, and coal).

5. Challenges and issues of bioLPG Europe context

The complexity of biomass and the problems associated with growing, harvesting, and transferring less dense feedstock to centralized bioreactors make n complex [32]. Apart from the logistical obstacle, more processing phases are still involved in transforming biomass into liquid fuel for transportation, such as hydrolysis, microbial fermentation, fuel separation, and pretreatment [33]. It can take longer than anticipated for lignocellulosic biofuels to reach the market due to these difficulties and a lack of government initiatives to generate demand for them. The synthesis of biofuels using feeds of lignocellulosic background, especially from algal biomass, presents more challenges regarding catalyst choice and hydrogen utilization [34].

Understanding the implications of the shift for their current business model and commercial, industrial, and agricultural enterprises is a problem as Europe moves toward a climate-neutral economy. Quick changes to the law may impact the demand for their goods and services, adding value to the switch to longer-term, lower-emission fuels for Europe's energy mix [35]. The lack of government initiatives to generate demand for bioLPG is also a significant concern. Different regulations apply to BioLPG in various EU members, leaving investors uncertain and making cross-border trademarks difficult. Differences in excise duties and carbon pricing mechanisms prevailing from country to country make BioLPG less competitive than fossil fuels. Furthermore, frequent changes in regulations and support schemes discourage long-term investment into BioLPG production infrastructure.

6. Result and conclusion

This chapter of the book titled "The Impact of BioLPGs on Carbon Footprint in a Context of European Countries" presents a comprehensive overview of the potential impact of BioLPGs on carbon footprint. We contend that there is an urgent requirement to transition to better sustainable bioeconomy techniques due to global worries about depleting the fossil fuel supply and the need for alternative fuels that generate fewer greenhouse gas emissions. The chapter highlights the potential of BioLPG as a renewable version of LPG made from biomass feedstocks such as rural buildups, ranger service waste, and waste cooking oil. BioLPG is a carbon-neutral alternative to traditional LPG and offers several benefits, including reduced net greenhouse gas emissions and lower carbon intensity.

The chapter evaluates the current state of research on BioLPG. It identifies critical challenges and issues, including the need for more research on the potential function of BioLPG in decarbonizing transportation. The authors also present potential solutions for broadly adopting BioLPG, such as utilizing multiple biomass feedstocks and maximizing waste value.

This chapter discusses the production process of BioLPG using commercial procedures employing hydrodeoxygenation (HDO) reactors, including the hydro treatment of vegetable oils (HVO) pathway, which involves hydrogenating the triglycerides contained in the raw vegetable material and yielding Propane [bioLPG] as a byproduct. We should mention that BioLPG's footprint varies and depends on the feedstock and situation cited in the European states.

In conclusion, the chapter presents BioLPG as a promising alternative to traditional LPG with several benefits, including reduced net greenhouse gas emissions and lower carbon intensity. We should mention that BioLPG is often eligible for government support through financial credits and meets biofuel requirements by EU countries under the Renewable Energy Directive. In general, the chapter gives a comprehensive outline of the potential impact of BioLPG on carbon footprint and highlights the need for sustainable bioeconomy techniques. We are adding value to the switch to longer-term, lower-emission fuels for Europe's energy mix. The lack of government initiatives to generate demand for bioLPG is also a significant concern. Different regulations apply to BioLPG in various EU members, leaving investors uncertain and making cross-border trademarks difficult. Differences in excise duties and carbon pricing mechanisms prevailing from country to country make BioLPG less competitive than fossil fuels. Furthermore, frequent changes in regulations and support schemes discourage long-term investment into BioLPG production infrastructure.

Author details

Amir Sepehri[1*] and Mohammad Kamal Ghassem Alaskari[2]

1 Research and Development of PGPZ, Tehran, Iran

2 PGPZ, Toronto, Canada

*Address all correspondence to: amirsepehry@rocketmail.com

IntechOpen

© 2024 The Author(s). Licensee IntechOpen. This chapter is distributed under the terms of the Creative Commons Attribution License (http://creativecommons.org/licenses/by/4.0), which permits unrestricted use, distribution, and reproduction in any medium, provided the original work is properly cited.

References

[1] Sanz-Hernández A, Esteban E, Garrido P. Transition to a bioeconomy: Perspectives from social sciences. Journal of Cleaner Production. 2019;**224**:107-119

[2] Amer M, Hoeven R, Kelly P, Faulkner M, Smith MH, Toogood HS, et al. Renewable and tuneable bio-LPG blends derived from amino acids. Biotechnology for Biofuels. 2020;**13**:125. DOI: 10.1186/s13068-020-01766-0

[3] Menezes NDA, Cunha ILC, dos Santos MT, Kulay L. Obtaining bio LPG via the HVO route in Brazil: A prospect study based on life cycle assessment approach. Sustainability. 2022;**14**(23):15734. Available from: https://www.mdpi.com/2071-1050/14/23/15734

[4] Lorenzi G, Baptista P, Venezia B, Silva C, Santarelli M. Use waste vegetable oil for hydrotreated vegetable oil production with high-temperature electrolysis as hydrogen source. Fuel. 2020;**278**:117991

[5] Iragavarapu GP, Imam SS, Sarkar O, Mohan SV, Chang YC, Reddy MV, et al. Bioprocessing of waste for renewable chemicals and fuels to promote bioeconomy. Energies. 2023;**16**(9):3873

[6] d'Amboise S, Mancarella A, Manelli A. Utilization of hydrotreated vegetable oil (HVO) in a euro 6 dual-loop EGR diesel engine: Behavior as a drop-In fuel and potentialities along calibration parameter sweeps. Energies. 2022;**15**(19):7202

[7] Johnson E. Process technologies and projects for BioLPG. Energies. 2019;**12**(2):250. Available from: https://www.mdpi.com/1996-1073/12/2/250/pdf

[8] Chen KC, Leach M, Black MJ, Tesfamichael M, Kemausuor F, Littlewood P, et al. BioLPG for clean cooking in sub-Saharan Africa: Present and future feasibility of technologies, feedstocks, enabling conditions and financing. Energies. 2021;**14**(13):3916

[9] Oh YK, Hwang KR, Kim C, Kim JR, Lee JS. Recent developments and key barriers to advanced biofuels: A short review. Bioresource Technology. 2018;**257**:320-333

[10] Materazzi M, Holt A. Experimental analysis and preliminary assessment of an integrated thermochemical process for production of low-molecular weight biofuels from municipal solid waste (MSW). Renewable Energy. 2019;**143**:663-678

[11] Srivastava A, Prasad R. Triglycerides-based diesel fuels. Renewable and Sustainable Energy Reviews. 2000;**4**(2):111-133

[12] Johnson E. A carbon footprint of HVO bio propane. Biofuels, Bioproducts and Biorefining. 2017;**11**(5):887-896

[13] Baldiraghi F, Stanislao MD, Faraci G, Perego C, Marker T, Gosling C, et al. Ecofining: New process for green diesel production from vegetable oil. Sustainable Industrial Chemistry. 2009:427-438

[14] Marker TL. Opportunities for Bio Renewables in Oil Refineries (No. DOEGO15085Final). Des Plaines, IL (United States): UOP LLC; 2005

[15] Islam MR, Ahmed ZU, Hossain KA. Numerical analysis of LOx-BioLPG combustion in high-pressure liquid rocket engine propulsion system.

South African Journal of Chemical Engineering. 2023;**45**:83-99

[16] Menezes NDA, Cunha ILC, dos Santos MT, Kulay L. Obtaining bioLPG via the HVO route in Brazil: A prospect study based on life cycle assessment approach. Sustainability. 2022;**14**(23):15734. Available from: https://www.mdpi.com/2071-1050/14/23/15734

[17] Unnasch S, Goyal L. Life Cycle Analysis of LPG Transportation Fuels under the Californian LCFS. Portola Valley, CA: Life Cycle Associates Report; 2017

[18] Rattenmaier N, Koppen S, Gärtner SO, Reinhardt G. Screening Life Cycle Assessment of Hydrotreated Jatropha Oil. Heidelberg, Germany: Institute for Energy and Environmental Research; 2008

[19] Arvidsson R, Persson S, Fröling M, Svanström M. Life cycle assessment of hydrotreated vegetable oil from rape, oil palm and Jatropha. Journal of Cleaner Production. Elsevier; 2011;**19**(2-3):129-137

[20] Garran D, Herrera I, Lechón Y, Lago C. Well-to-tank environmental analysis of a renewable diesel fuel from vegetable oil through co-processing in a hydrotreatment unit. Biomass and Bioenergy. Elsevier; 2014;**63**:239-249

[21] Blonk Agri Footprint. Agri - footprint 2.0. Gouda, The Netherlands: Blonk Agri Footprint; 2015

[22] Fritsche UR, Sims RE, Monti A. Direct and indirect land-use competition issues for energy crops and their sustainable production–an overview. Biofuels, Bioproducts and Biorefining. Wiley; 2010;**4**(6):692-704

[23] UK Department of Energy & Climate Change, Atlantic Consulting and Menecon Consulting. RHI Evidence Report: Bio-Propane for Grid Injection. London: Department of Energy & Climate Change; 2014

[24] de Jong P, Torres EA, de Melo SABV, Mendes-Santana D, Pontes KV. Socio-economic and environmental aspects of bio-LPG and bio-dimethyl ether (bio-DME) production and usage in developing countries: The case of Brazil. Cleaner and Circular Bioeconomy. 2023;**6**:100055. Elsevier

[25] Behrends FJ. Greenhouse Gas Footprint of Biodiesel Production from Used Cooking Oils [Thesis]. CS Utrecht: Utrecht University; 2018

[26] Price R. "Clean" Cooking Energy in Uganda–Technologies, Impacts, and Key Barriers and Enablers to Market Acceleration. Brighton, UK: Roz Price Institute of Development Studies; 2017

[27] Sunde K, Brekke A, Solberg B. Environmental impacts and costs of hydrotreated vegetable oils, transesterified lipids and woody BTL—A review. Energies Journal. MDPI; 2011;**4**(6):845-877

[28] Eu I. Renewable energy directive. Koeln, Germany: ISCC System GmbH Hohenzollernring; 2018. Available from: https://certificates.iscc-system.org/cert-pdf/EU-ISCC-Cert-DE119-60171375.pdf

[29] Johnson E. Disagreement over carbon footprints: A comparison of electric and LPG forklifts. Energy Policy. 2008;**36**(4):1569-1573. Elsevier

[30] Røyne F, Peñaloza D, Sandin G, Berlin J, Svanström M. Climate impact assessment in life cycle assessments of forest products: Implications of

method choice for results and decision-making. Journal of Cleaner Production. 2016;**116**:90-99. Elsevier

[31] van Grinsven A. Emissions of (Bio) LPG and Other Energy Carriers in Domestic Heating, BBQs and Forklift Trucks. Delft the Netherlands: CE Delft Oude Delft; 2021

[32] Balan V. Current challenges in commercially producing biofuels from lignocellulosic biomass. ISRN biotechnology. Hindawi International Scholarly Research Notices; 2014;**2014**:463074. DOI: 10.1155/2014/463074

[33] Xiu S, Zhang B, Shahbazi A. Biorefinery Processes for Biomass Conversion to Liquid Fuel. London, UK: Intech open; 2011. pp. 167-190. DOI: 10.5772/16417

[34] Furimsky E. Hydroprocessing challenges in biofuels production. Catalysis Today. 2013;**217**:13-56

[35] WLPGA, Report. The role of LPG and LPG in Europe. The-2019-Report. WLPGA. 2019. Available from: https://www.wlpga.org/wp-content/uploads/2020/03/The-Role-of-LPG-Bio-LPG-in-Europe-The-2019-Report.pdf